Grade Inflation

Springer
New York
Berlin
Heidelberg
Hong Kong
London
Milan
Paris
Tokyo

Grade Inflation

A Crisis in College Education

Valen E. Johnson

 Springer

Valen E. Johnson
Department of Biostatistics
UM School of Public Health, Bldg II
1420 Washington Heights
Ann Arbor, MI 48109
valenj@umich.edu

Library of Congress Cataloging-in-Publication Data
Johnson, Valen E.
    Grade inflation : a crisis in college education / Valen E. Johnson.
      p.   cm.
    Includes bibliographical references and index.
    ISBN 0-387-00125-5 (softcover : alk. paper)
    1. College students—Rating of—United States.   2. Grading and marking
(Students)—United States.   3. Student evaluation of teachers—United States.
I. Title
LB2368 .J65   2003
378.1'67—dc21   2002042741

ISBN 0-387-00125-5   Printed on acid-free paper.

Text design by Steven Pisano.

Printed in the United States of America.

9 8 7 6 5 4 3 2 1   SPIN 10898123

www.springer-ny.com

Springer-Verlag   New York Berlin Heidelberg
A member of BertelsmannSpringer Science+Business Media GmbH

Acknowledgments

T his book examines the phenomenon of grade inflation and its impact on postsecondary education. Much of this investigation is based on research conducted by others, but a substantial portion originates from data collected at Duke University during the 1998–99 academic year. As a consequence, one might be tempted to conclude that grade inflation is a more serious problem at Duke than it is elsewhere. This is, of course, not true.

Grade inflation is a national, if not international, problem. Recent scandals over grading practices at Ivy League colleges and other top universities illustrate this point all too well. That this book was written using data collected at Duke University should therefore not be regarded as an indictment of Duke, but should instead be viewed as an indication that many professors and administrators at Duke were concerned with this problem and were willing to confront it. This book would not have been possible without their support, and probably would not have been written at many other universities.

Many individuals deserve credit for their role in facilitating campus-wide discussions of grading at Duke and for making the DUET experiment a reality. Among these are Professor Daniel Graham, who chaired a committee that focused attention on the need to reform grading practices, and Professor Daniel Gauthier, who also served on this committee and later helped implement the DUET experiment. Professors Angela O'Rand and John Richards, former chairs of the Arts and Sciences Council and Academic

Affairs Committee, respectively, also played pivotal roles in this process. Angela was an ardent supporter of the DUET experiment and managed to delay its demise for at least one semester longer than I thought possible. John's support was essential in gaining approval for the experiment from the Academic Affairs Committee. Support from Provost John Strohbehn and his successor, Peter Lange, was also critical. Deans Richard White, Robert Thompson, and Mary Nijhout helped guide the project during its nascent stage and assisted in the construction of the DUET survey instrument. Harry DeMik provided important advice regarding the availability and limitations of data from the registrar's office. Ben Kennedy, a student government leader at Duke, along with fellow students Jeff Horowitz and Tammy Katz, was instrumental in encouraging students to participate in the DUET experiment and, in so doing, greatly enhanced the value of data collected. I also thank my colleagues in the Institute of Statistics and Decision Sciences, and in particular its former director Mike West, for their support of the project. Finally, I would like to thank Dalene Stangl, Peter Mueller, Daniel Gauthier, John Kimmel, and my wife Pamela Johnson for many helpful comments that improved the presentation of material in this book.

Contents

1 Introduction

ON MAY 13, 1997, DUKE UNIVERSITY'S ARTS AND Sciences Council rejected a proposal to change the way students' grade point averages (GPAs) are computed. The proposal, based on a statistical adjustment scheme called the achievement index, was intended to correct student GPAs for differences between the grading policies employed by different professors and departments, and to provide feedback to faculty concerning their individual grading practices. Of the 61 members of the council, 19 voted against and 14 voted in favor of the proposal. I suspect the proposal would have fared better if the ACC basketball tournament hadn't started that afternoon. Humanities faculty really aren't into Duke basketball the way science faculty are.

The impetus for the achievement index proposal had been provided some months before by the university provost, who had requested that the faculty convene a committee—later to be known as the Committee on Grades—to investigate the problem of grade inflation. Over the previous decade, grades at Duke University had risen steadily until in 1997, over 45% of all grades awarded to undergraduates at Duke were A's of one flavor or another. Fewer than 15% of grades were C+ or lower. Of course, grade inflation was not unique to Duke. Nearly every other university in the United States had experienced a similar trend during the 1990s, and in fact, Duke was barely keeping pace with its "peer institutions." For example, the mean GPA at Dartmouth in 1994 had risen to 3.23, up from 3.06 in 1968. To combat grade inflation there, the Dartmouth faculty adopted a plan in the fall of 1994 to list median course grade and class size next to student grades on university transcripts. Hints as to the success of this strategy are reflected in a later article by Bradford Wilson, who reported that in the fall of 1999 the proportion of Dartmouth grades that were A's or A−'s had risen to 44%. Meanwhile, the median GPA at Princeton in 1997 was

3.42, and at Harvard, 46% of grades awarded to undergraduates in the 1996–1997 academic year were A's or A–'s [Wil99].

In convening the faculty committee to investigate the problem of grade inflation, Provost Strohbehn, a recent transfer from Dartmouth, had probably hoped that the committee would propose a plan similar to the one implemented at his former institution. However, the chair of the Duke committee, Professor Daniel Graham, took a somewhat broader view and led the committee through a comprehensive investigation of the problems associated with Duke's grading policies. In the end, the conclusion of this committee was that disparities in grading practices, rather than grade inflation, were responsible for most of the problems usually associated with lenient grading. The achievement index was proposed as an attempt to remedy these ills.

Harvey Mansfield, a professor of government at Harvard University and a controversial opponent of inflationary grading practices, writing in an editorial on this topic, states, "In a healthy university, it would not be necessary to say what is wrong with grade inflation. But once the evil becomes routine, people can no longer see it for what it is. Even though educators should instinctively understand why grade inflation is a problem, one has to be explicit about it" [Man01, B24]. When I and other members of the Committee on Grades made our proposal to the Duke University Arts and Sciences Council, we made the unfortunate assumption that we were operating in a "healthy university." What Mansfield, members of the Committee on Grades, many members of the Duke faculty, and the general public did not or do not understand are the deep philosophical divisions that exist between members of the academic community regarding the meaning and purposes of student grades.

Most members of the Committee on Grades subscribed to what might be called the traditional view of grading. From this perspective, grades are devices used by faculty to maintain academic standards and to provide summaries of student progress. Grade inflation undermines both of these goals. William Cole, an instructor in Romance languages and literatures at Harvard University, summarized the traditionalists' concerns over the relationship between grade inflation and the lessening of academic standards as follows:

In many courses, faculty members are giving out relatively high grades for average or subpar work. While such inflation might look innocent, it has in fact grown into a significant problem, with no end in sight. By rewarding mediocrity we discourage excellence. Many students who work hard at the outset of their college careers, in pursuit of good grades and honors degrees, throw up their hands upon seeing their peers do equally well despite putting in far less effort.

Today's students, it seems to me, are highly pragmatic individuals who, while eager to learn, are even more eager to succeed. After all, one does not get into a highly competitive private college—or any other selective institution—in the first place without figuring out the prerequisites for admission and then managing to fulfill them. But if, after being admitted, a student sees that all that we demand for success is minimal effort, that's all we'll get [Col93, B1].

The impact of grade inflation on the faculty's ability to evaluate students is no less severe. Mansfield put it thus:

Grade inflation compresses all grades at the top, making it difficult to discriminate the best from the very good, the very good from the good, the good from the mediocre. Surely a teacher wants to mark the few best students with a grade that distinguishes them from all the rest in the top quarter, but at Harvard that's not possible. Some of my colleagues say that all you have to do to interpret inflated grades is to recalibrate them in your mind so that a B+ equals a C, and so forth. But the compression at the top of the scale does not permit the gradation that you need to rate students accurately [Man01, B24].

As the data presented earlier attest, compression at the high end of the grading scale is not unique to Harvard. Indeed, many universities have been forced to adopt a new grade to overcome the lack of an appropriate grade to indicate truly outstanding performance: the A+. Thirty years ago, an A+ was a grade awarded in elementary schools. It usually wasn't a grade given at the high school level, and was certainly not issued at most colleges. But many colleges now include the A+ as a valid grade. Duke does, and now the number of A+'s awarded at Duke is also on the rise.

Recent trends in grade inflation at universities across the nation suggest that most faculty reject the traditional definition of student grades, or at least choose not to adhere to it. Still, many professors are willing to accept the traditional premise that a grade of C should represent average work, a grade of B above average work, and so on. But for one reason or another, they insist that their students are above average. Interestingly, the reference group to which students are compared varies. Sometimes it consists of students at other universities; at other times, students in other classes at the same university; or, in the case of Duke, the alumni.

The comparison with Duke alumni arose during the discussion of the achievement index. When confronted with the fact that the average GPA had increased by about one-third of a grade during the 1990s, several faculty members pointed out that the average SAT scores of Duke students had also risen during that period. If student quality had improved, they argued, why shouldn't average grades also increase?

Similar arguments have been forwarded by professors at Dartmouth. Bemoaning the failure of Dartmouth undergraduates to gain acceptance at top graduate schools in the 1960s, the Dartmouth faculty intentionally revised their grading system so as to improve their students' odds. According to Noel Perrin, a professor of environmental studies at Dartmouth, the faculty at Dartmouth

> ...began systematically to inflate grades, so that our graduates would have more A's to wave around. But, if you preferred to be accurate, you would say that we simply recognized and began to follow the trend toward national standards, national reputations, and national comparison groups. No longer do most of us on the faculty just compare one Dartmouth student with another; we take into account the vast pool of college students nationwide, all five million of them. That is, we imagine our students at a mythical Average U., and give the grades they would get there [Per98, A68].

At Duke, such reasoning is also used by many instructors to justify course grades that are unusually high by departmental or university standards. The most frequent beneficiaries of this rationale are students in seminar courses and courses that require special talents, like music or art classes. In such cases, faculty members maintain that their students are superior to students who did not register for their class. As a consequence of some unobserved course selection mechanism, they insist that their high marks are not only warranted, but are mandated out of a sense of fairness.

For these professors, the problem with the traditional perspective on grading is that it fails to account for variation in student quality from class to class or school to school. Perrin's assertion that a B at Dartmouth is equal to an A at most other colleges is probably correct. As he states in his article, the average SAT verbal score of a Dartmouth student was over 700 in 1998, undoubtedly higher than the average verbal SAT score at Average U. To counter this problem, schools that have held the line on grade inflation, like Swarthmore, Reed, and the University of the South, have enacted policies to inform potential employers and graduate schools of their stricter grading standards. For example, the University of the South now adds students' percentile rankings to their transcripts. Reed College includes a statement of their grading policy with transcripts, and Swarthmore maintains contact with professional schools in order to keep them apprised of its current grading policies [Com99].

It goes without saying that traditionalists spurn the idea that better students are the cause of grade inflation. Mansfield summarizes the objection as follows:

> At Harvard, we have lost the notion of an average student. By that I mean a Harvard average, not a comparison with the high-school average that enabled our students to be admitted here. When bright students take a step up and find themselves with other bright students, they should face a new, higher standard of excellence....
>
> Some say Harvard students are better these days and deserve higher grades. But if they are in some measures better, the proper response is to raise our standards and demand more of our students. Cars are better-made now than they used to be. So when buying a car, would you be satisfied with one that was as good as they used to be [Man01, B24]?

I wonder also how many of the professors who make the argument that their students are better than students at other universities would assign uniformly low grades if fate had landed them at Below Average U.? And should community colleges and lower-rung state schools really be prevented from assigning A's?

In their recent report on university assessment practices, Rosovsky and Hartley describe yet another defense of grade inflation. They note that a portion of the academy now ascribe to

> ...the view that low grades discourage students and frustrate their progress. Some contend it is defensible to give a student a higher grade than he or she deserves in order to motivate those who are anxious or poorly prepared by their earlier secondary school experiences. Advocates of this opinion contend that students ought to be encouraged to learn and that grades can distort that process by motivating students to compete only for grades.... A more radical view holds that it is inappropriate for a professor to perform the assessment function because it violates the relationship that should exist between a faculty member and students engaged in the collaborative process of inquiry. Some critics of grades argue that it is a distorting, harsh, and punitive practice.

They go on to point out, however, that

> ...grades certainly are not harsh for those who do well, and empirical evidence for the hypothesis that lowering the anxiety over grades leads to better learning is weak...Although the rejection of grading does not represent the academic mainstream, the criticisms are influential in some circles [RH02, 3–4]....

A more cogent alternative to the traditional perspective on grading practices is the postmodern view. From this perspective, science and the scientific method, observation of natural phenomena, and objective consideration of evidence are replaced by, or at least supplemented with, a critical assessment of the scientist and the inherent biases that accompany his membership in "dominant groups." An objective view of reality and search for truth is replaced by an emphasis on divergent representations of reality. Gone also are notions of hierarchies in which the relative values of ideas and knowledge systems are compared.

With the loss of objectivity, a postmodernist is much less likely to assign poor grades. In her explanation of the postmodern perspective on teaching and grading, Diana Bilimoria states this point as follows:

> Teachers' increasing awareness of the biases inherent in modern science is likely to affect their evaluations of students' acquisition of subject matter. Because disciplinary content domains are increasingly open to diverse interpretations and the inclusion of alternative representations, the scope of what is legitimate and appropriate knowledge in the academic enterprise is widened. The global questioning of tenets once held to be singularly true allows a larger number of students to display with greater diversity a legitimate and appropriate grasp of a widened content. Consequently, grade distributions are higher than they were before the advent of postmodern challenges....
>
> As postmodern perspectives gain greater legitimacy, teachers' openness to different conclusions and more diverse methods at arriving at them favor higher grade distributions because evaluation criteria are broadened and there are many, rather than a few, acceptable discourses in which students can engage. Students are empowered to challenge not just the insights but also the methods presented by dominant orthodoxies. Failure to display reason, analysis, objective consideration of evidence, and distance is much less used as an explanation for poor grades, as these keystones of modern science are themselves shown to be biased in favor of certain, but not other, views, and are hence no more valid than any other method of arriving at conclusions [Bil95, 443].

From a traditionalist's perspective (like mine), this postmodern view of grading seems bizarre. In practice, however, many professors are more comfortable with this perspective on grading—or less extreme versions of it—than they are with the traditional interpretation of grades. And therein lies the rub. Within any college or university, several very different approaches to grading are used to evaluate students.

The disparities in grading practices that result from these divergent views of grading have serious consequences, the most obvious being inequitable assessment of students. Clearly, students who take classes from faculty who grade leniently have a better chance of finishing college with higher GPAs—and thus better career prospects—than do students who take most of their classes with instructors who grade more stringently.

Students respond to this reality in two ways. First, they preferentially enroll in classes with instructors who grade leniently. Second, they provide more favorable course evaluations to these instructors.

The impact on faculty members is equally severe. Stringent graders, by virtue of their lower course enrollments and lower course evaluations, are less likely to receive tenure, salary increases, and promotions. Professors know this and respond by raising their grades to meet student expectations. Grade inflation ensues when stringently grading professors chase their more leniently grading colleagues toward the beginning of the alphabet. It is exacerbated when students differentially choose courses with the winners, since there are then more students taking classes with instructors who assign "above average" grades. Finally, with traditional incentives for students to achieve eliminated, academic standards fall.

Why, then, have such inequities in grading practices become accepted? The thesis of this book is that grading inequities persist because their consequences are misunderstood. In particular, they exist and are perpetuated by the following myths:

1. Student grades do not bias student evaluations of teaching.
2. Student evaluations of teaching provide reliable measures of instructional effectiveness.
3. High course grades imply high levels of student achievement.
4. Student course selection decisions are unaffected by expected grading practices.
5. Grades assigned in unregulated academic environments have a consistent and objective meaning across classes, departments, and institutions.

Ironically, the last myth is often advocated most fervently by individuals who, in most other aspects of their professional lives, reject the notion of objective, quantifiable, and hierarchical measures of quality.

The first myth—concerning the relationship between the grades students get from their instructors and the ratings that students give to their instructors—is simultaneously the most studied and most misrepresented topic in educational assessment. This relationship also plays a central role in determining the dynamics of inflationary grading practices. For this reason, two chapters are devoted to its examination. Although these chapters are quite detailed, they serve to establish a context from which the results of later chapters can be gauged.

The first chapter to investigate the effects of student grades on student ratings of instruction, Chapter 3, begins with a review of previous observational studies. Generally speaking, these studies focus on the computation and interpretation of empirically observed correlations between student grades and student evaluations of teaching. Over sixty such studies are summarized in this chapter. A preponderance of these report positive correlations between student grades and student evaluations of teaching.

Next, several of the more prominent theories that have been advanced to explain the generally observed positive association between grades and student evaluations of teaching are explored. These theories range from the teacher-effectiveness theory at one end of the spectrum to the grade-leniency theory at the other. The former is based on the notion that good teaching leads to both higher student grades and higher ratings of instruction, while the latter basically asserts that students simply reward leniently grading instructors with higher course evaluations.

In addition to observational studies, Chapter 3 also reviews a number of experiments in which student grades were systematically manipulated in an attempt to establish more clearly the nature of the link between grades and student evaluations of teaching. Each of these experiments appears to confirm a causal link between higher student grades or grade expectations and higher student evaluations of teaching. However, such findings have posed a serious threat to educational researchers, who in many instances have had a vested interest in demonstrating the validity of the teacher–course evaluation instruments necessary for the conduct of their research. For this reason, and others, the methods employed in these experiments have been subject to extensive criticism. Whether this criticism has been justified is not entirely clear, but an interesting insight into this debate was provided by Anthony Greenwald, who noted that "although these reservations deserve serious consideration, it must be noted also that the standard strategy for opposing published experiments on methodological grounds—repeating the experiments with improved methods—was never pursued by any of the critics of the grade-manipulation studies.... Although the conclusions of these experiments have been questioned by critics, those conclusions have not been empirically refuted" [Gre97, 1183]. Chapter 3 concludes by reviewing the more prominent of these experiments and summarizes the criticisms and conclusions drawn from them.

In Chapter 4, new evidence concerning the nature of the relationship between student grades and student evaluations of teaching is presented. The source of this evidence derives from a novel web-based survey instrument called DUET that was used at Duke University during the 1998–1999 academic year. Two analyses based on data from this survey are detailed. In the first, student evaluations were regressed on student grades, mean item responses collected from other students, and prior student interest variables. This analysis proves particularly useful for assessing the validity of each of the major theories described in Chapter 3 for explaining observed associations between grades and teacher–course evaluations. Second, survey responses collected from students before they received their final course grades are compared to their responses after they received their final course grades. This analysis provides compelling evidence of a causal effect of student grade expectations on student evaluations of teaching.

The DUET experiment also plays a pivotal role in several of the analyses presented in later chapters. In addition to providing insights into the relationship between student grades and student evaluations of teaching, data collected during this experiment also provide a fresh look into many of the assessment mechanisms that influence undergraduate behavior. The major innovations of this experiment were the use of web-based technology to collect and disseminate course evaluation data from and to students, the linking of course evaluation data to student records, and the observation of students as they acquired and utilized information from the website. The conduct of the DUET experiment and the unique opportunities it provides for studying student behavior are summarized in Chapter 2.

The utility of student evaluations of teaching for studying educational processes is examined in Chapter 5. Emphasis in this chapter focuses on the validity of these forms in predicting student learning and achievement. Alternative measures of the validity of teacher–course evaluation forms, and the motivation for these alternative measures, are also discussed. In related analyses, the value of DUET survey items, student background variables, and course characteristics for predicting student learning are examined. An important feature of these analyses involves the measure of student learning employed as the dependent variable. Unlike most previous research in educational assessment, in which outcome measures have been defined in terms of performance of students on standardized end-of-term examinations, the outcome measure employed in the analyses of Chapter 5 is based on student performance in follow-on courses.

In Chapter 6, focus shifts from the impact of assessment practices on faculty to the effects of grading practices on students. In particular, the influences of expected grading policies on student course enrollment decisions are examined. Two approaches are described for investigating this issue. In one, the probability that a student takes a second course in a department is modeled as a function of the student's grades in the first course taken in that field [SWL91]. The second approach probes the influence that students' prior knowledge of a course's mean grade has on their decisions to enroll in future offerings of the same course. In this analysis, students' prior knowledge of mean course grades is gleaned from records of student queries of mean course grades to the DUET website. The conclusions from both analyses are that grading practices have a significant impact on the courses that students elect to take.

Finally, in Chapter 7, the myth that grades have an objective interpretation that transcends departments and instructors is dispelled. Incontrovertible evidence is provided that grading practices do differ by academic field, and, in concordance with commonly held perceptions, tend to be most lenient in the humanities and most stringent in the natural sciences, mathematical sciences, and economics. Case studies are presented to illustrate the potential gravity of these effects on the career paths of students. To correct for these disparities in grading practices, two grade adjustment methods are presented. The statistical properties of these adjustment schemes are found to be similar despite minor variations in their treatment of atypical courses. Both methods have the potential for broad application in America's higher educational system, with uses ranging from the adjustment of grades reported on official transcripts, to determination of honors distinctions, to imposition of constraints on mean course grades, to simply providing feedback to instructors regarding their grading practices. A summary of findings and recommendations for reform are presented in the concluding chapter.

SUMMARY

G rade inflation and methods for reversing it have been discussed at nearly every American college at one time or another during the past decade. In a few instances, these discussions have led to actual reforms. But in most cases, the debate has been successfully quelled by opponents of reform who invoke one or more of the myths listed in the previous section.

Regardless of one's philosophy toward grading, it must be acknowledged that assigning higher-than-average grades is, at the very least, convenient. In many cases it is career enhancing. It should therefore come as no surprise that reforms that entail more uniform grading are often opposed by a majority of a university's faculty. Furthermore, the very culture of academia makes reform difficult. Generally speaking, about one-half of any university's faculty will be adversely affected by reform measures, and every member of the other half will have a better idea on how reform should be implemented! Reaching a consensus on both the need and mechanism for reform is therefore a daunting task.

The purpose of this book is twofold. First, I hope to expose many of the myths associated with grade inflation, the use of student evaluations of teaching for administrative reviews of faculty, the effects of disparate grading practices on students, and the effects of grading practices on student enrollment patterns. In particular, I provide evidence that higher grades do lead to better course evaluations, student course evaluations are not very good indicators of how much students have learned, higher mean course grades do not reflect higher levels of student achievement, and students can (and probably do) manipulate their GPAs by selecting courses with instructors who grade leniently. Second, I hope to expand the discussion of college grading from the sensationalized topic of grade inflation to the broader issue of how assessment practices can be modified to reflect student and faculty achievement more fairly. Several strategies for accomplishing this goal are described at the end of the book.

In reading this book, it should be borne in mind that for ethical reasons, it has not generally been possible for researchers to perform experiments that directly measure the effects of grading policies on student learning and behavior. Because of such ethical constraints, relatively sophisticated statistical analyses have been required to extract information from those data that are available. Although I have attempted to minimize the description of the technical detail involved in these analyses, some level of detail was necessary in order to substantiate results. However, for the convenience of those readers who are interested in results and are willing to accept heuristic explanations for how the analyses were performed, most technical details have been relegated to appendices. Also included in the appendices are detailed discussions of how nonresponse (i.e., the fact that students self-selected for the DUET survey) may have affected study conclusions. This material may be passed over on first reading without loss of continuity.

The DUET Experiment

An online course evaluation experiment called DUET was conducted at Duke University during the 1998–1999 academic year. This experiment possessed several unique features, including the solicitation of survey responses from students both before and after they had received their final course grades and monitoring of students as they viewed mean grades and course evaluations of courses taught in the past.

D EBATE OF THE GRADING-REFORM PROPOSAL among members of the Duke faculty spawned numerous questions regarding the role of grades at the university. Issues of particular concern included the effects that grades had on student course selection decisions and student responses to end-of-term course evaluations, and the extent to which GPA reflected a fair summary of students' academic performance. Although the achievement index proposal was defeated, many faculty members were not satisfied that these underlying issues had been adequately addressed. A short time later, I was asked by Professor John Richards, chair of the Academic Affairs Committee, and Robert Thompson, dean of Trinity College of Arts and Sciences, to chair a committee to continue an examination of these issues. Other individuals asked to join the committee included Professor Daniel Gauthier, from physics; Mary Nijhout, associate dean of Trinity College; Benjamin Kennedy, vice president of academic affairs in the Duke student government; and Jeff Horowitz, cofounder and a coadministrator of a student website called Devilnet. In addition to collecting information regarding the effects of grading practices on undergraduate education, the committee was charged with studying the feasibility of collecting end-of-term course evaluations electronically over the web.

Shortly after its formation, the members of the committee agreed to proceed with the construction of a website that would have as its primary purpose the collection and dissemination of course evaluation data. As a secondary objective, the website would serve as an experimental platform to collect data useful for answering some of the questions that had been posed during the grading-reform debate.

After several months of meetings in which items to be included on the course evaluation form were negotiated among committee members, followed by a perfunctory review of the experimental design by a human subjects protocol committee and numerous meetings with the Academic Affairs Committee to gain final approval to proceed with the experiment, the DUET (Duke Undergraduates Evaluate Teaching) website was launched in the fall of 1998. It was scheduled to run for three years, through the spring of 2001, but for reasons detailed later in this chapter, it was terminated after the spring survey period in 1999. The basic operation of the website is described below.

First, to publicize the DUET website and to encourage student participation, Dean Thompson agreed to send personalized letters to all full-time Duke undergraduates requesting their participation in an experiment designed to study the feasibility of collecting course evaluation data over the web. Included in this letter were individual student access codes that permitted students to enter the website. Ideally, student passwords from the university computing system would have been used in place of these codes, but it proved technically impossible to transfer these passwords to Webslingerz Inc., the company that administered the site, in time for the launch of the survey. Students were also informed that they would be able to review both course evaluation data entered by other students and mean course grades of courses taught in the past after they had completed their surveys.

The DUET website was activated for two three-week periods, the first beginning the week prior to fall registration in 1998 and the second a week prior to spring registration in 1999. Both periods fell approximately ten weeks into their respective semesters. The survey was conducted immediately before and during registration as an incentive for students to participate; by completing the DUET survey prior to registration, students could view course evaluation data and grade data for courses they planned to take the following semester. With the exception of first-year students in the fall of 1998, all students completed the survey for courses they had taken the previous semester. As part of the experimental design, first-year students were asked to complete the survey for courses they were currently taking. A timeline for the DUET experiment is depicted in Figure 1.

Upon entering the DUET website, students were initially confronted with text informing them that course evaluation data col-

FIGURE 1

DUET timeline.

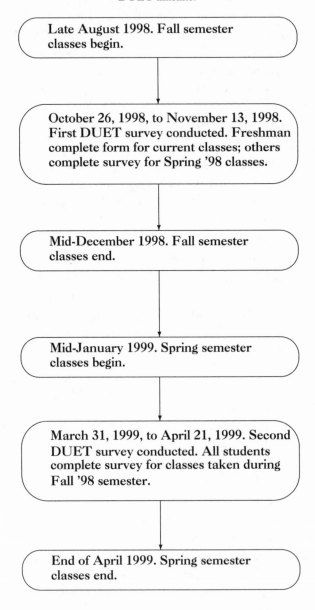

lected on the site would be used as part of a study investigating the feasibility of collecting course evaluation on the web. They were also told that their responses would not be accessible to either their instructors, other members of the faculty, other students, or university administrators. After indicating their consent to participate in the study, items on the survey were presented to students in groups of 5–7 questions. Each course they had taken the previous semester (or, for freshmen in the fall, that they were currently taking) appeared simultaneously as a row in a table, while item responses were listed in columns next to each course. Item text appeared above the courses, and students were required to click a button in response to each item for every course before the next set of items appeared. With only one or two exceptions, all items included a "Not Applicable" response. After completing the 38 items on the survey, students were invited to view course evaluation data collected from other students, course mean grades for all classes that had been taught in the past five years, and their adjusted GPAs. They were also encouraged to express free-form opinions on grades and grade adjustments. Survey items and possible responses are listed below.

DUET SURVEY ITEMS

1. In what format was this course primarily taught?
   1) Lecture 2) Seminar 3) Project
   4) Combination of Lecture, Seminar, and Project
   5) Skills course 6) Not Applicable

2. Did you take this course to satisfy a
   1) Major requirement 2) Distributional requirement (not major requirement) 3) Elective

3. How would you rate your interest in the subject matter of this course prior to enrolling in it?
   1) Very Low 2) Low 3) Moderate 4) High 5) Very High
   6) Not Applicable

4. What proportion of the class sessions did you find challenging?

1) less than 35% 2) 35-65% 3) 65-85% 4)` 85-95%
5) More than 95% 6) Not Applicable

5. What proportion of the class sessions did you find
   relevant to course objectives?
   1) less than 35% 2) 35-65% 3) 65-85% 4) 85-95%
   5) More than 95% 6) Not Applicable

6. What proportion of the reading and writing assignments
   did you find challenging?
   1) less than 35% 2) 35-65% 3) 65-85% 4) 85-95%
   5) More than 95% 6) Not Applicable

7. What proportion of the reading and writing assignments
   did you find relevant to the course material?
   1) less than 35% 2) 35-65% 3) 65-85% 4) 85-95%
   5) More than 95% 6) Not Applicable

8. Do you know what the goals of the course were?
   1) No 2) Yes 3) Not Applicable

9. Approximately how many hours per week did you spend
   in this course attending class meetings, discussion
   sessions, and labs?
   1) Less than 2 hours 2) 2-3 hours 3) 3-5 hours
   4) 5-8 hours 5) More than 8 hours 6) Not Applicable

10. Approximately how many hours per week did you spend
    outside of class completing reading and written
    assignments, studying for exams, and completing
    projects, etc.?
    1) Less than 2 hours 2) 2-3 hours 3) 3-5 hours
    4) 5-8 hours 5) More than 8 hours 6) Not Applicable

11. What proportion of the reading assignments did you
    complete?
    1) less than 30% 2) 30-50% 3) 50-75% 4) 75-90%
    5) More than 90% 6) Not Applicable

12. What proportion of the written assignments did you
    complete?
    1) less than 50% 2) 50-75% 3) 75-90% 4) More than 90%
    5) Not Applicable

13. What proportion of the classes did you attend?
    1) less than 30% 2) 30-50% 3) 50-75% 4) 75-90%
    5) More than 90% 6) Not Applicable

14. How difficult was the material taught in this course
    compared to other courses that you've taken at Duke?
    1) Not difficult 2) Less difficult than average
    3) Average 4) More difficult than average
    5) Very difficult 6) Not Applicable

15. How effective was the instructor in encouraging students
    to ask questions and express their viewpoints?
    1) Very poor 2) Poor 3) Fair 4) Good 5) Very Good
    6) Excellent 7) Not Applicable

16. Did this class have a TA (teaching assistant)?
    1) Yes 2) No 3) Not Applicable

17. How important was the TA's role in the course compared
    to that of the primary instructor(s)?
    1) Not important
    2) Less important than the primary instructor(s)
    3) Equally important as primary instructor(s)
    4) More important than the primary instructor(s)
    5) Not Applicable

18. How would you rate the TA of this course?
    1) Very bad 2) Bad 3) Fair 4) Good 5) Excellent
    6) Not Applicable

19. How effective were exams, quizzes, and written
    assignments at measuring your knowledge of course
    material?
    1) Very bad 2) Bad 3) Fair 4) Good 5) Excellent
    6) Not Applicable

20. How was this class graded?
    1) Very leniently 2) More leniently than average
    3) About average 4) More severely than average
    5) Very severely 6) Not Applicable

21. How valuable was feedback on examinations and graded
    materials?
    1) Very poor 2) Poor 3) Fair 4) Good 5) Very Good
    6) Excellent 7) Not Applicable

22. What grade do you expect (or did you get) in this class?
    1) A+
    2) A
    3) A-
    4) B+

5) B
6) B-
7) C+
8) C
9) C-
10) D+/D/D-
11) F
12) Not Applicable

23. On a scale from 1 to 5, with 1 being "completely unaware"
and 5 being "completely aware", how aware were you of
how this course would be graded when you enrolled in it?
1) 1
2) 2
3) 3
4) 4
5) 5
6) Not Applicable

24. How much did your knowledge of how the course would be
graded positively affect your decision to enroll in it?
1) No effect or negative effect 2) Slight effect
3) Moderate effect 4) Significant effect
5) Very significant effect 6) Not Applicable

25. How well did you learn and understand the course
material?
1) Very poorly 2) Poorly 3) Fair 4) Well 5) Very Well
6) Not Applicable

26. How much did you learn in this course compared to all
courses that you have taken at Duke?
1) Much less than average 2) Less than average
3) Average 4) More than average
5) Much more than average 6) Not Applicable

27. How would you rate the instructor(s) knowledge of course
material?
1) Very bad 2) Bad 3) Fair 4) Good 5) Very Good
6) Excellent 7) Not Applicable

28. How easy was it to meet with the instructor outside of
class?
1) Very difficult 2) Difficult 3) Not Hard 4) Easy
5) Very Easy 6) Not Applicable

29. How would you rate the organization of the instructor(s)
in this course?

1) Very poor 2) Poor 3) Fair 4) Good 5) Very Good
6) Excellent 7) Not Applicable

30. How good was the instructor at relating course material
   to current research in the field?
   1) Very bad 2) Bad 3) Fair 4) Good 5) Very Good
   6) Excellent 7) Not Applicable

31. To what extent did this instructor demand critical or
   original thinking?
   1) Never 2) Seldom 3) Sometimes 4) Often 5) Always
   6) Not Applicable

32. The instructor's concern for the progress of individual
   students was
   1) Very Poor 2) Poor 3) Fair 4) Good 5) Very Good
   6) Excellent 7) Not Applicable

33. How would you rate this instructor's enthusiasm in
   teaching this course?
   1) Very bad 2) Bad 3) Fair 4) Good 5) Very Good
   6) Excellent 7) Not Applicable

34. How good was the instructor(s) at communicating course
   material?
   1) Very bad 2) Bad 3) Fair 4) Good 5) Very Good
   6) Excellent 7) Not Applicable

35. How does this instructor(s) compare to all instructors
   that you have had at Duke?
   1) Very bad 2) Bad 3) Fair 4) Good 5) Very Good
   6) Excellent 7) Not Applicable

36. How would you rate your interest in the subject matter
   covered in this course now?
   1) Very Low 2) Low 3) Moderate 4) High 5) Very High
   6) Not Applicable

37. Would you recommend this course to others?
   1) Yes 2) No 3) Not Applicable

38. Would you take another course from this instructor?
   1) Yes 2) No 3) Not Applicable

Several aspects of the DUET website should be emphasized
here since they play a prominent role in later analyses. First, all
course-evaluation data collected during the experiment were linked

to respondents. Because complete student transcripts were also available from the university registrar, this meant that student grades and other background variables could be used in conjunction with the evaluation data to provide a clearer indication of the meaning of student responses to the survey. Some of the background variables that were available for this purpose included student major, courses taken and grades received, SAT math and verbal scores, high school GPA, gender, and ethnic group.

Second, and unbeknownst to students, a record was kept of every student query into the course-evaluation and mean-grade databases. That is, every time a student viewed either a mean course grade or a histogram summary of data entered by other students, the date, time, course, and instructor of that course, along with the querying student's ID, were all appended to a second database. These data were later used to investigate the effect that knowledge of mean grades and course evaluations had on students' course enrollment decisions.

A final experimental aspect of the DUET website involves the survey design for first-year students. First-year students completed the survey for their fall courses twice, once before completing their fall courses and receiving their final grades, and once after. Because one DUET survey item asked students what grade they received or expected to receive in their courses, the responses collected from freshman at the two time points provide an ideal mechanism for investigating the influence that expected and received student grades have on student evaluations of teaching. Analyses based on these data are presented in the last sections of Chapter 4.

A number of factors that affected student participation in the survey are also worth mentioning here. The first concerns a technical problem that surfaced shortly after the DUET website was launched. This problem involved limits on the volume of traffic that the server hosting the site could handle and resulted in students being locked out of the site—often after already completing a substantial portion of their survey—whenever traffic volume exceeded the server's capacity. The host was eventually replaced by a server that could handle more connections, but in the interim many students' surveys were lost, and many students decided that participation in the study was not worth the hassle.

A similar problem centered on the use of non-Duke passwords for system access. As mentioned previously, it proved to be administratively impossible for students to use their Duke computing account passwords to access the site. Instead, students had to use passwords that were included in the mailing from Dean Thompson soliciting their participation in the experiment. Unfortunately, many students lost this correspondence before the study began, and so were unable to access the site without first requesting their password from the site administrator.

As serious as the impact on survey response from these technical glitches was, faculty opposition to the site soon proved to be far more threatening. Faculty opposition to the experiment was caused by several factors, the most prominent being a perception held by many professors that posting student evaluations of their courses on the web represented a violation of their privacy. And to be truthful, I must confess that had the experiment been initiated by someone else, I would have had similar concerns given the academic climate that then prevailed at Duke. During that period, professors had almost complete control over their course evaluations, at least as to whether they were released to students and other faculty members. In fact, student efforts to publish course synopses were regularly hampered by instructors who refused even to allow representatives of the Duke student government from visiting their offices to compile summary statistics from their evaluations.

Other members of the faculty were suspicious of any use of electronic media to collect information. Such concerns ranged from healthy skepticism over the confidentiality of student responses to something approaching paranoia. For example, one professor circulated an email in which she stated that the on-line course evaluation system was "far too close to techniques of surveillance for my comfort," and likened the DUET website to intelligence gathering activities of "the US military (and other military units around the world)," who were tasked "to design, refine, and implement research and development efforts that will give them superiority over information."[1]

---

[1] Personal email communication.

In addition to these complaints, which were, to some extent, anticipated, there was also substantial and unexpected opposition to the DUET experiment from the Department of Mathematics. Some of this resistance likely stemmed from the privacy concerns mentioned above, with more generated by a rather vocal member of the mathematics department who had received particularly bad ratings for an introductory calculus course he had taught the previous semester. But most seemed to arise from a fear among many in the math department that their course enrollments would suffer if mean grades of previous courses continued to be displayed on the web. They apparently anticipated one of the major conclusions of the study, though they were likely naive in thinking that they were not already suffering from exactly this phenomenon. Further opposition from the math department was ignited when another member of the department was asked by a student to explain why the class mean grade in his class was an entire letter grade lower than the mean class grades awarded in three other sections of the same course. As it happens, the mean GPAs and SAT scores for students in all sections of this course were nearly identical, so the student may have had a point. In any case, the senior professor in question was not amused by the inquiry.

In April of 1999 I received a letter written on behalf of five departments (Asian and African Languages, Art and Art History, Germanic Languages and Literature, Romance Studies, and Slavic Languages and Literatures) and 28 faculty members (21 from Mathematics) demanding that their courses and departments be excluded from the DUET website. In this letter they threatened me with unspecified "further action" if their request was not satisfied. Copies of this letter were sent to the provost, two deans, and several other administrative officers. I suspect that even greater pressure was applied to members of the administration to end the study, and shortly thereafter, Dean Thompson succumbed to the inevitable and asked me to terminate the experiment after the spring 1999 collection period. Several weeks later, the Arts and Sciences Council considered a motion to postpone the experiment until issues regarding the use of faculty members as possible subjects in the experiment and privacy concerns had been addressed. A vote on the motion was postponed until the following fall, by which time the plug had already been pulled.

For many of the analyses planned using data collected during the DUET experiment, the early termination of the experiment was not particularly problematic. During the two semesters that the DUET website was active, 11,521 complete course evaluations were recorded, with each evaluation consisting of 38 item responses for a single course. Of the 6,471 eligible full-time, degree-seeking students who matriculated at Duke University after 1995, 1,894 students (29%) participated in the experiment. Of these, approximately one-half participated in both the fall and spring surveys. Participation during the following year would likely not have increased greatly, and response patterns for different student groups would probably also have remained about the same. There was, however, one analysis that was severely affected by the experiment's early termination: an analysis of the effects of student grade expectations on course enrollments. For that analysis, data that would have been collected during the third semester of the experiment were crucial for assessing the impact of sample selection effects. To overcome the absence of these data, I decided to conduct a follow-on email survey of selected study participants during the 1999–2000 academic year. Details of this follow-on survey and its role in subsequent analyses are described in Chapter 6.

Histogram summaries of student responses collected during the experiment are displayed in Figure 2.

APPENDIX: ISSUES OF NONRESPONSE

Because only 29% of eligible students participated in the DUET experiment, it is important to consider the effects of "nonresponse." Survey nonresponse occurs when a subset of a study population fails to participate in a survey. If students' decisions to participate in the DUET survey were strongly related to the way they (would have) responded to the survey, then the generalizability of study conclusions based on survey data collected only from those students who did participate is severely compromised. On the other hand, if the "response mechanism" is uncorrelated or

FIGURE 2

Histogram summary of DUET responses. For clarity, "Not Applicable" responses were excluded from the plots.

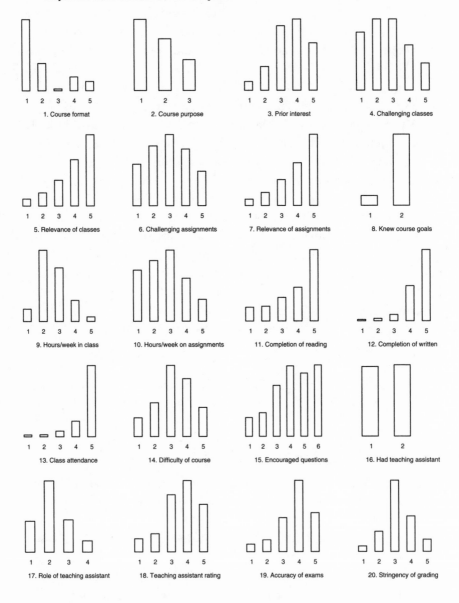

FIGURE 2 (*continued*)

21. Usefulness of exams

22. Expected grade

23. Prior knowledge of grading

24. Grading affected enrollment

25. Learned course material

26. Comparative learning

27. Instructor knowledge

28. Instructor availability

29. Instructor organization

30. Related course to research

31. Critical thinking

32. Instructor concern

33. Instructor enthusiasm

34. Instructor communication

35. Instructor compared to others

36. End interest in subject

37. Recommend course to other

38. Another course from instructor

only moderately correlated with student response patterns, then nonresponse does not seriously limit the scope of study conclusions. Unfortunately, examining differences between the way study participants did respond to the survey and the way nonparticipants may have responded to the survey is difficult, owing to the absence of data from the latter group.

Students' decisions to abstain from participation in the DUET experiment may be attributed to several causes, including time constraints experienced by students during the survey periods, unavailability of a convenient access port to the DUET website, privacy concerns, objections to the perceived purpose of the study, and, of course, student apathy. As mentioned above, technical difficulties at the website and problems associated with the use of student access codes may also have contributed to the low response rate. For instance, problems with the database server caused an almost constant stream of service interruptions during the first three days of the initial survey period. As a result of these system failures, numerous student responses to the survey were lost, and it is difficult to estimate the number of students who may have permanently dropped out of the experiment due to frustration encountered in accessing the site during the initial solicitation.

In addition to early technical problems with the server, the fact that students could not use their usual computer passwords to access the site also reduced participation. Over one hundred requests for access codes were received by the webmaster during the two survey periods. Although access codes were provided to all students from whom such requests were received, subsequent participation rates among these students—along with an unknown number of other students who gave up without requesting access codes—were, without doubt, substantially reduced.

Although these technical problems undoubtedly affected survey participation, it is unlikely that they introduced significant response biases. However, other causes of nonresponse may have been less benign. For example, women were less likely to participate in the survey than men were, and African-American students were less likely to participate than Caucasians. To better understand the effects of these and other sources of nonresponse, the relationships between an individual's propensity to participate in the survey and an individual's demographic attributes were examined.

Several demographic classifications that might plausibly be associated with an individual's decision to participate in the survey were extracted from data provided by the Office of the University Registrar. These classifications included gender, ethnicity (Asian, African-American, Hispanic, Native American, White, and Other), college major and school, academic year, and GPA.

As mentioned above, gender was a highly significant predictor of survey participation. One-third of eligible males (33%) participated in the study, while only 25% of females did. So too, were differences associated with ethnic groups. Participation among students who identified themselves as African-American was 14%, while for students who identified themselves as Native American participation was 25%. Participation rates for other ethnic categories were 38% for Asian students, 26% for Hispanic students, and 29% for White students. Students who declined to place themselves into one of these categories participated with 35% probability. Clearly, if mechanisms responsible for differences in response rates among different gender and ethnic categories are also linked to differences in response patterns to the DUET items, then adjustments would be necessary to correct for the disproportionate numbers of students from each group who participated in the experiment.

To evaluate the impact of these demographic classifications on response patterns in the survey data, histogram estimates of the probabilities that each gender and ethnic group responded to each item were constructed. These histograms are displayed in Figures 3 and 4.

Figure 3 shows that item responses did not vary substantively with gender. The greatest proportional differences between the responses of women and men occurred on the first two items, which queried the format and reason for taking a course. These differences are probably attributable to tendencies for women and men to major in different academic fields, and the concomitant differences in the types and numbers of courses required in these different

FIGURE 3

Comparison of DUET responses for women and men. For each response category, women's responses are depicted in the left bar and men's responses are depicted in the right bar.

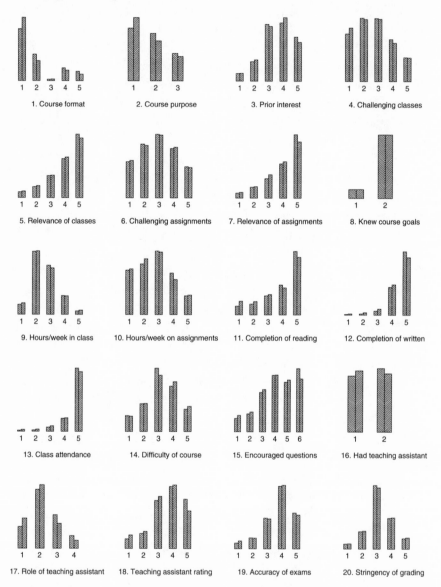

FIGURE 3 *(continued)*

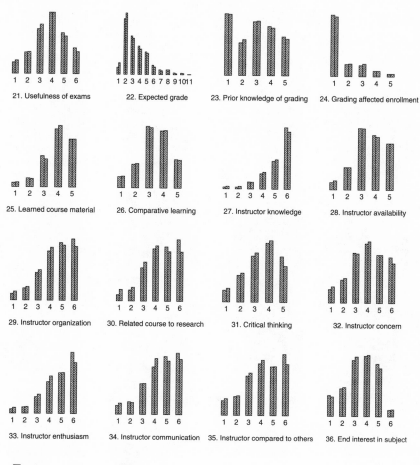

21. Usefulness of exams

22. Expected grade

23. Prior knowledge of grading

24. Grading affected enrollment

25. Learned course material

26. Comparative learning

27. Instructor knowledge

28. Instructor availability

29. Instructor organization

30. Related course to research

31. Critical thinking

32. Instructor concern

33. Instructor enthusiasm

34. Instructor communication

35. Instructor compared to others

36. End interest in subject

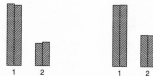

37. Recommend course to other 38. Another course from instructor

FIGURE 4

Comparison of DUET responses by ethnic group. For each response category
and item, the order of the vertical bars is White, Asian, African-American,
Hispanic, Native American, and Unknown.

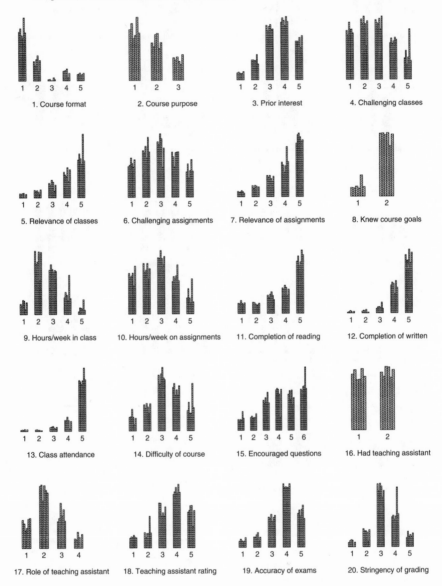

FIGURE 4 (*continued*)

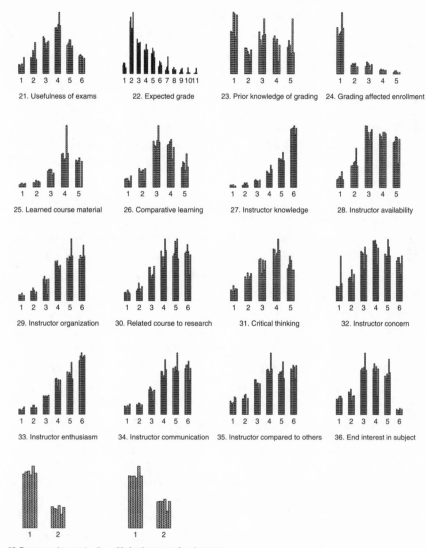

fields. Aside from these items, it is evident from Figure 3 that females were slightly more likely to find courses and assignments relevant, to complete written and reading assignments, and to give slightly higher ratings on items that evaluated specific teaching traits. Whether these differences can be attributed to gender differences or differences in the types of courses taken is not immediately clear.

A more quantitative summary of the effects of gender nonresponse on the distribution of observed responses can be constructed by comparing the proportion of responses actually collected in each item category to the estimated proportion of responses that would have been collected had women and men responded to the survey with equal probability. Using demographic data obtained from the registrar's office to make this correction, the probability that the "same" response would be drawn from both the observed and "nonresponse-corrected" distributions exceeded 98% for all items on the survey.[2] In practical terms, then, differences between the response rates of women and men probably had no substantive impact on survey conclusions.

Differences in the response patterns between ethnic groups are more pronounced than they are for gender. The most discrepant cells in Figure 4 correspond to respondents who classified themselves as Native Americans (the fifth vertical bar in each response category). However, because only 20 individuals identified themselves in this category, these deviations may simply represent random fluctuations in responses obtained from this group. Also, because this group represented only a small proportion of the eligible student population, corrections to response totals to account for these differences would not have a significant effect on survey totals. Similarly, the comparatively low response rates obtained from African-American students, though disappointing, does not constitute a significant threat to the veracity of survey conclusions, since only 8.3% of eligible students fell into this category. Furthermore, response patterns obtained from African-American students did

---

[2]Technically, the value of 98% was calculated by subtracting the total variation distance between the distributions from 1.0. It thus represents the probability that the same value would be drawn from both distributions in γ-coupling.

not differ greatly from responses obtained from other non–Native American ethnic categories.

The quantitative statistic cited above for gender differences extends also to differences in response probabilities attributable to ethnic groups. In this case, the probability that the same response would be drawn from the distribution actually observed for each item as from the distribution corrected for differences in response probabilities between ethnic groups exceeds 99% for every survey item. As was the case for gender, nonresponse patterns attributable to ethnic group do not appear problematic.

In addition to gender and ethnicity, variation in response patterns and response rates according to educational factors like student GPA, academic division of students' majors, and academic year were examined. Figure 5 displays the proportion of responses received from students who had GPAs less than 3.3 (left bar) against responses from students who had GPAs greater than 3.3. Once again, the response patterns do not vary substantially across the different demographic groups. In this case, the only notable differences between the responses collected from students having high GPAs and low GPAs were observed for item 11 (proportion of reading assignments completed), item 13 (class attendance), and item 22 (expected grade). For each of these items, differences between the distributions of responses from the two groups lean in the expected direction. The response rates for the two groups were 32% for students who had GPAs less than 3.3, and 27% for students with GPAs greater than 3.3.

The probability that equivalent responses would be drawn from the observed distribution and the distribution corrected for this GPA categorization exceeds 97% for all items except item 22, which probed the grade that students expected or had received in a course. Differences in the response patterns for this item are, of course, expected. Effects of nonresponse on study conclusions attributable to student GPA are discussed at appropriate points in the sequel.

Figure 6 depicts histogram estimates of responses by academic division. Academic divisions represented in this plot are, from left to right, Social Sciences, Engineering, Humanities, and Natural Sciences. According to this figure, response patterns for these groups

FIGURE 5

Comparison of DUET responses by GPA. For each response category and item, the vertical bars on the left represent responses obtained from students whose GPAs were less than 3.3, while the bars on the right depict responses obtained from students having GPAs greater than 3.3.

FIGURE 5 *(continued)*

21. Usefulness of exams

22. Expected grade

23. Prior knowledge of grading

24. Grading affected enrollment

25. Learned course material

26. Comparative learning

27. Instructor knowledge

28. Instructor availability

29. Instructor organization

30. Related course to research

31. Critical thinking

32. Instructor concern

33. Instructor enthusiasm

34. Instructor communication

35. Instructor compared to others

36. End interest in subject

37. Recommend course to other

38. Another course from instructor

FIGURE 6

Comparison of DUET responses by academic division. For each response category and item, the order of the vertical bars is Social Sciences, Engineering, Humanities, and Natural Sciences.

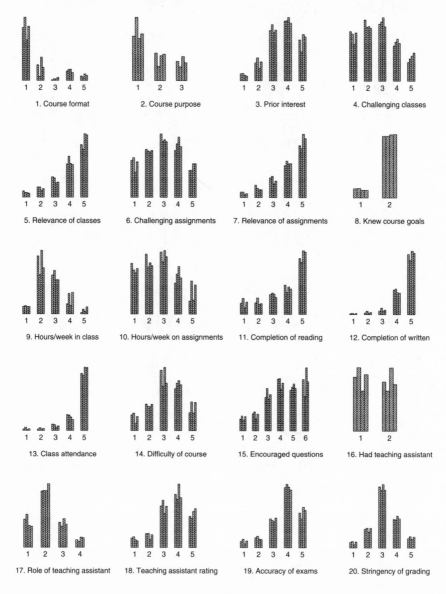

1. Course format
2. Course purpose
3. Prior interest
4. Challenging classes
5. Relevance of classes
6. Challenging assignments
7. Relevance of assignments
8. Knew course goals
9. Hours/week in class
10. Hours/week on assignments
11. Completion of reading
12. Completion of written
13. Class attendance
14. Difficulty of course
15. Encouraged questions
16. Had teaching assistant
17. Role of teaching assistant
18. Teaching assistant rating
19. Accuracy of exams
20. Stringency of grading

FIGURE 6 *(continued)*

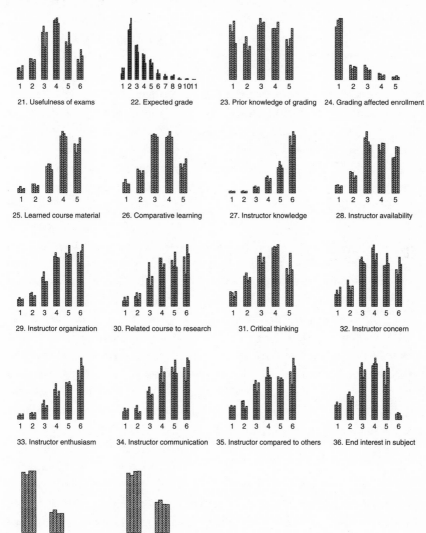

21. Usefulness of exams

22. Expected grade

23. Prior knowledge of grading

24. Grading affected enrollment

25. Learned course material

26. Comparative learning

27. Instructor knowledge

28. Instructor availability

29. Instructor organization

30. Related course to research

31. Critical thinking

32. Instructor concern

33. Instructor enthusiasm

34. Instructor communication

35. Instructor compared to others

36. End interest in subject

37. Recommend course to other

38. Another course from instructor

are less consistent than were the patterns observed for groupings based on gender and ethnicity.

Engineering students tended to give higher responses on the items that probed hours/week on assignments (item 10), but lower responses to items probing prior subject matter interest (item 3), the extent to which instructors encouraged questions (item 15), useful-ness of exams (item 21), prior knowledge of course grading policies (item 23), instructor knowledge (item 27), instructor organization (item 29), instructor's ability to relate course material to current research (item 30), instructor enthusiasm (item 33), and instructor communication (item 34). The response rates for engineers were also unusually high—39%—compared to the other groups, although they represented only a relatively small proportion of students eligible for the survey (15%, compared to 21% for humanities, 23% for natural and mathematical sciences, and 42% for social sciences). Because of the relatively large differences between the response patterns of students from the School of Engineering and those from Arts and Sciences, and their overrepresentation in the study sample, several later analyses were conducted without data collected from engineering students. In particular, several of the analyses conducted in Chapter 5 are limited to responses received only from Arts and Sciences students. Differences in the response patterns of social science, humanities, and natural science students were less significant.

Despite the differences between engineering students and non-engineering students, the observed distribution of responses and the distribution of responses corrected for the differential participation rates of students from different divisions were again quite similar. In this case, the probability of drawing the same response from both distributions exceeded 97% for all survey items.

Student responses broken out by academic year are depicted in Figure 7. Except for item 23, which probed student knowledge of course grading policies prior to enrollment, the response patterns by year group are also surprisingly similar. In the case of item 23, first-year students' responses were much higher in the first category, cor-responding to little or no prior knowledge of grading policy. Because freshmen participants enrolled for their first-semester courses prior

FIGURE 7

Comparison of DUET responses by academic year. The order of the response proportions are, from left to right, seniors, juniors, sophomores, and freshmen.

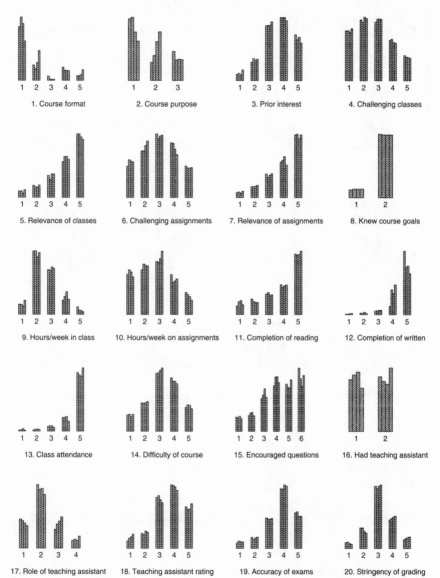

FIGURE 7 *(continued)*

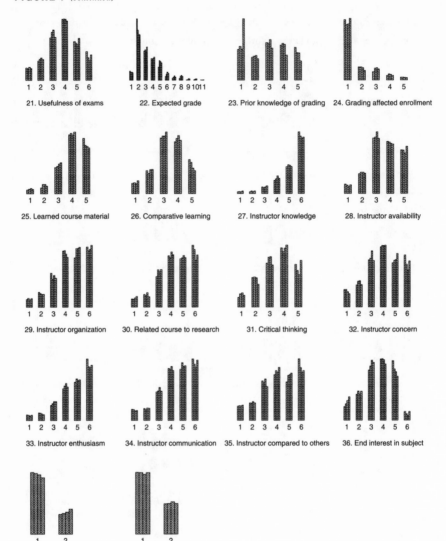

21. Usefulness of exams

22. Expected grade

23. Prior knowledge of grading

24. Grading affected enrollment

25. Learned course material

26. Comparative learning

27. Instructor knowledge

28. Instructor availability

29. Instructor organization

30. Related course to research

31. Critical thinking

32. Instructor concern

33. Instructor enthusiasm

34. Instructor communication

35. Instructor compared to others

36. End interest in subject

37. Recommend course to other

38. Another course from instructor

to arriving at Duke, the discrepancy for first-year students on item 23 is expected.

Despite the similarity of response patterns by year, it is important to note that the probability that a student participated in the experiment did vary significantly with academic year. Only 15% of fourth-year students completed the DUET survey, while 27% of third-year students did. Among second- and first-year students, participation rates were 39% and 35%, respectively.

Differences in response rates between year groups can probably be attributed to "intellectual dropout." As a class of students moves closer to graduation, it seems that an increasing proportion of students lose interest in academics. Typically, these students either turn their attention to life after college or focus on nonacademic activities. The decreasing trend in the participation rates of students by academic year may simply be a consequence of increasing apathy. Conversely, the similarity of responses obtained from student participants from different academic years can probably be attributed to the fact that those students who did choose to participate in the survey were the same students who also chose to remain academically engaged.

Nonresponse attributable to student apathy or intellectual dropout does present a potential threat to the generalizability of survey conclusions. Because nonacademically oriented students were probably underrepresented in the experiment, and because there is no clear measure of students' academic motivation in records provided by the registrar, corrections for this response mechanism cannot be made. Intuitively, however, it seems reasonable to assume that less academically oriented students are probably swayed more by grading considerations when choosing their classes and rating their instructors than are more serious students. Thus, in terms of impact on study conclusions, this source of response bias likely means that conclusions based on the DUET data are *conservative* in the sense that the measured effects of grades on student course selection and student evaluations of teaching would have been more pronounced if the sample had been more representative of the student body at large.

Of course, the conservative effect of student apathy and intellectual dropout on nonresponse must be balanced against the possibility that those students most interested in selecting courses according to grade distributions were overrepresented in the survey sample. Although the letter that solicited student participation on the survey was worded so as to emphasize the pedagogical benefits of collecting teacher–course evaluation data on-line, the fact that mean course grades could be viewed on the DUET website was also prominently featured. Whether this effect was more important than the opposing effect of student apathy and intellectual dropout is an open question. However, even if grades had absolutely no influence on the behavior of every student who chose not to participate in the survey, the impact of grades on students as measured by the DUET data is still highly significant; study conclusions based on the DUET data unquestionably apply to nearly 30% of all Duke undergraduates, and nearly 40% of first- and second-year students.

Finally, the plots and quantitative summaries discussed in this appendix suggest that nonresponse associated with known demographic variables can largely be ignored. Effects of other sources of nonresponse are more difficult to assess, and arguments for both positive and negative biases on study conclusions are plausible. However, until compelling arguments for biases in either direction can be substantiated, it seems reasonable to analyze survey responses as they were collected, and to regard conclusions based on these analyses as being approximately representative of conclusions that would have been drawn using a less self-selected survey sample.

3 Grades and Student Evaluations of Teaching

For eighty years, researchers have investigated the relation between grades and student evaluations of teaching. Nearly all studies conducted have resulted in reports of a positive correlation between these variables, but debate continues over the cause of this association. This chapter reviews past studies and experiments that have probed this relationship and summarizes several of the more common explanations for it.

T O THE UNINITIATED, THE NATURE OF THE RELATION between grades awarded by professors to their students and the ratings awarded by students to their professors is intuitively obvious: Instructors who grade leniently are more likely to gain approval of their students, have better rapport in the classroom, and be reviewed positively on course evaluation forms. This is not to say that grades alone determine how students rate their professors, only that higher grades have an effect, and the effect tends to be positive.

The view from within academia is somewhat different. Many professors and administrators are convinced that student evaluations of teaching (SETs) are either uncorrelated or negatively correlated with student grades. Surprisingly, this view finds substantial support in the educational research literature. Indeed, several of the earliest and most often cited studies of this relationship support this assertion, and confusion over the nature of the relationship between grades and student evaluations of teaching has led to an extended debate among educational researchers that continues today. Although most researchers now accept the proposition that assigned grades are positively correlated with student evaluations of teaching, many of these same individuals do not believe that this correlation represents a "biasing" effect. Instead, they, like many professors and administrators, attribute the positive correlation between grades and teaching evaluations to one of several alternative explanations. The most common of these is the teacher-effectiveness theory (e.g., [GG97, McK79]).

The teacher-effectiveness theory is based on the supposition that students learn more in courses taught by effective teachers. Because they learn more, students are rewarded with higher grades, and the resulting positive correlation between grades and teacher–course evaluations is not only not the result of an underlying bias, but is a desirable feature of student evaluations of teaching. This theory was championed in many of the earlier studies that examined the relation between grades and student evaluations of teaching (e.g., [CGM71, Ray68, Rem30, RR72, Sch75, VF60]).

The antipode of the teacher-effectiveness theory is the grade-leniency or grade-satisfaction theory. The basic tenet of this theory is that students reward instructors who reward them. According to this theory, positive correlations between mean course grades and student evaluations of teaching represent a bias. It is a bias in the sense that the effect of grading on teaching evaluations represents a factor not related to either effective teaching or student learning.

The grade-leniency theory gained credence in the 1970s through a series of experiments and correlational studies (see [Fel76] for a review), but has been largely ignored in recent years due, at least in part, to a quest by educational psychologists to design survey instruments that measure "true" dimensions of teaching effectiveness. Because grading biases diminish the validity of these forms, researchers have sought alternative interpretations of the relation between grades and student evaluations of teaching, generally to the tune that grading biases are not biases after all.

Greenwald and Gilmore [GG97] discuss several less extreme explanations for the observed correlation between student grades and student evaluations of teaching and suggest that the operation of the mechanisms underlying each of these theories can be discerned through an examination of observed patterns of student grades versus student evaluations of teaching. The manifestation of these patterns in the DUET data is discussed in the next chapter.

Regardless of the true relationship between grades and teaching evaluations, the fact that many instructors perceive a positive correlation between assigned grades and student evaluations of teaching has important consequences when there also exists a perception that student course evaluations play a prominent role in promotion, salary, and tenure decisions. As Winsor states, "This perceptual construct adds incentive to acquiesce to apparent student wishes regarding quantity and quality of academic excellence. If giving relatively high evaluations is seen as a means to further a career as an educator, and if these evaluations are viewed partly as a popularity contest where relevant demands of academic rigor are negatively valued by students, the psychological pressures to lessen requirements and to escalate rewards appears an obvious result" [Win77, 83].

The extent to which this perception has affected American universities is difficult to assess, although the existence of this effect is almost beyond question. Students often do complain loudly and vigorously about grades that they feel are undeservedly low, and there is little doubt that the same students frequently register their dissatisfaction with their grades by denigrating teaching effectiveness on end-of-course teaching evaluations. The perception by professors that student course evaluations play a prominent role in promotion, salary, and tenure decisions is well founded; a vast majority of American universities now include some summary of these evaluations when making promotion and tenure decisions.

To avoid the perils described by Winsor, it seems clear that student evaluations of teaching must be adjusted for the biasing effects of instructor grading policies, at least to the extent that these effects are indeed biasing. Furthermore, the form of this adjustment must be made known to instructors, and instructors must be convinced that the adjustment eliminates the potential benefits that might otherwise be accrued through lenient grading. Unfortunately, there is currently no consensus regarding the existence or extent of the bias associated with the effects of grading policies on student evaluations of teaching, and so it is not surprising that few, if any, universities make systematic adjustments to student evaluations of teaching to account for assigned grades.[1]

In the remainder of this chapter, results from previous observational and experimental studies are presented, and each of the major theories proposed to explain the correlation between grades and teaching evaluations is examined in light of the evidence provided by these studies. Together with evidence derived from the DUET experiment and presented in the next chapter, these studies provide convincing evidence that a grading bias in student evaluations of teaching does exist. The existence of this bias highlights the

---

[1]A notable exception is the University of Washington; information about the adjustment schemes adopted there may be found at www.washington.edu/oea/uwrepts.htm.

need to adjust faculty teaching evaluations prior to their use in administrative reviews.

OBSERVATIONAL STUDIES

C oncern over correlations between student grades and student evaluations of teaching has a long history, dating back at least seventy years to a study reported by Remmers in 1928 in the journal *School and Society* [Rem28]. Since Remmers' article, literally hundreds of observational studies have been conducted to investigate this relationship. Most of these studies have concluded that a positive correlation between either received or expected grades and student evaluations of teaching does exist, even if reported results have been discrepant both in the mode and magnitude of this association.

Observational (or correlational) studies[2] of the relation between grades and course evaluations are generally grouped according to whether data were recorded at the level of the individual student or aggregated at the classroom level. Data collected at the student level can, in theory, be used to assess the relationship between the performance of students in individual classes and their ratings of teacher effectiveness. Data aggregated by class are usually considered best for assessing the effects of instructor leniency on teaching evaluations, since such data can be used to compare average class grades and average course evaluations.

Tables 1 and 2 provide summaries of observational studies that investigated the relationship between student grades and student evaluations of teaching. Data in these tables are based largely on comprehensive reviews compiled by Feldman [Fel76] and Stumpf

[2]Observational studies are studies in which researchers do not (intentionally) modify the environment or system they are investigating. In contrast, experimental studies involve manipulation of environmental variables, usually for the purpose of establishing a causal relationship between variables. Because observational studies do not involve such manipulations and are unable to control for various unobserved factors that may affect the values of recorded variables, establishing causality on the basis of observational studies is usually difficult.

TABLE 1

Student-Level Studies of Grades and SETs. Values reported in this table represent correlations between grades (expected or received) and student evaluations of teaching effectiveness for studies in which students were considered the unit of analysis. Correlations subscripted with a 1 are based on additional analyses or correspondence with the cited authors as reported in [SF79]. Statistically significant ($p < 0.05$) and highly significant ($p < .01$) correlations are indicated by $*$ or $**$. A $+$ or $++$ in the correlation column indicates a statistically significant positive regression coefficient in studies for which no partial correlation is reported.

Author(s)	Correlation	Comments
Barnoski & Sockloff [BS76]	.10** to .16**	Correlation between expected grade reported on questionnaire and four factors thought to measure instructor and class effectiveness
Blass [Bla74]	−.10 to .60**	Correlations between student reported midterm grades and nine dimensions of teaching; eight of the nine dimensions resulted in positive correlations, with six of these eight being significant at either the 0.05 or 0.01 level
Blum [Blu36]	−.05, .04	Correlation between actual and expected grades in two classes; correlations computed as if all students were enrolled in a single class
Caffrey [Caf69]	.23** to .32**	Correlations of actual course grade and several dimensions of teaching effectiveness
Centra [Cen77]	−.15 to .92**	Correlations of grades on common final exam and several dimensions of teaching effectiveness for two courses

TABLE 1 *(continued)*

Author(s)	Correlation	Comments
Centra & Linn [CL73]	.44**	Correlations of expected grade with global teaching effectiveness variable
Domino [Dom71]	.18, .39**	Correlation with evaluation of course and teacher effectiveness, respectively
Endo & Della-Piana [EDP76]	−.04 to .13	Correlations based on student's final exam grade (unknown to student) on seven dimensions of teaching effectiveness; reported correlations were averaged across five teachers, with only a few highly significant within class correlations reported
Frey [Fre73]	.14* to .91**	Correlations based on final examination scores, corrected for student SAT score, and six dimensions of teaching effectiveness; averaged across 15 sections of calculus and multivariate calculus with students completing the SET after students had received their final course grades
Frey [Fre76]	.19*	Mean correlation of six dimensions of teaching effectiveness and final exam performance, corrected for SAT score; SET completed after receipt of final course grades by mail-in survey

TABLE 1 *(continued)*

Author(s)	Correlation	Comments
Garverick & Carter [GC62]	$.21_1$*	Correlation with instructor effectiveness and expected grade
Granzin & Painter [GP73]	$.16$**(.09), $.21$**(.15**)	Correlation of expected (actual) grade with instructor and overall course, respectively
Gigliotti & Buchtel [GB90]	$.15$** to $.30$**	Correlation of six dimensions of teaching effectiveness on actual course grade; *lower* but positive correlations reported with expected grade and grade discrepancy
Hocking [Hoc76]	$.13_1$	Multiple regression on change of instructor effectiveness and student interest variables on change in expected grade, measured at middle and end of term; reported t-values between 0.195 and 3.53**
Holmes [Hol71]	$.10_1$**	Highly significant F-statistics reported for 28% of 126 one-way analyses of variance, each of which examined the effect of actual grade (A, B, or C) on 1 of 18 evaluation items administered to seven classes
Hudelson [Hud51]	$.19$**	Correlation of actual grade and instructor rating
Kapel [Kap74]	$.16$** to $.19$**	Correlation of expected grade to several dimensions of teaching effectiveness
Kelley [Kel72]	$.22_1$	Multiple regression of course evaluation and professor evaluation on expected grade

TABLE 1 (*continued*)

Author(s)	Correlation	Comments
		and 17 additional concomitant variables
Kovaks & Kapel [KK76]	.28**	Correlation between expected grade and overall measure of teaching effectiveness
Miller [Mil72]	.05*	Correlation with expected grade
Pascale [Pas79]		Evaluated effect of knowledge of final grade (previous exam results known) on course evaluations; statistically significant differences found only on items involving feedback to students with no direct examination of effects of expected or actual grade received
Owie [Owi85]	*	Significant F-statistic in one-way analysis of variance in which difference in expected and obtained grades were considered treatment effect on global teacher evaluation score
Peterson & Cooper [PC80]	.44*	Correlation between expected grade and overall evaluation of teaching with a similar correlation found between the self-assigned grade of ungraded students and the same evaluation score
Pratt & Pratt [PP76]	.11, .25*	Correlations with received and expected grade and overall instructor evaluation; students given input into weighting assigned to different components of final grade

TABLE 1 *(continued)*

Author(s)	Correlation	Comments
Remmers [Rem30]	.07	Average correlation across several dimensions of teaching, eleven instructors, and 17 classes
Rosenshine, Cohen, & Furst [RCF73]	.27**, .09**, .12**	Correlation of expected grade with course rating, instructor rating, and whether class was worthwhile to attend
Russell & Bendig [RB53]	$.20_1$*	One-way analysis of variance between groups defined by differences between predicted and obtained grades and global ratings of instructors
Scheurich, Graham, & Drolette [SGD83]		No significant effect of expected grade on overall instructor evaluation found in multiple regression across all students and within 31 class groups; significant, simple correlations of .25 and .23 between expected grade and overall teacher evaluation were reported
Schwab [Sch75]	.06* to .11*	Partial correlations of expected grade with five dimensions of teaching effectiveness
Schuh & Crivelli [SC73]	$.48_1$**	One-way analysis of variance of midterm grade on overall student rating of teacher
Seiver [Sei83]	+	Reported significant regression effect of expected grade on global evaluation of teacher in ordinary least squares regression; no significant relation was found for simultaneous, two-stage least

TABLE 1 *(continued)*

Author(s)	Correlation	Comments
		squares regression when grade was also regressed on faculty evaluation and other variables
Spencer [Spe68]	.3	Correlation between overall value of course and grade
Spencer & Dick [SD65]	.85**	Correlation between "attitude toward instructor" and average grade for 160 students divided into 10 sections of a speech class at Penn State University
Starrak [Sta34]	.15*	Correlation of actual grade and mean course rating
Treffinger & Feldhusen [TF70]	.11	Correlation with final grade in course
Voeks & French [VF60]	$.05_1$	Correlation between grade distribution and global student ratings of student-selected faculty members

and Freedman [SF79]. Table 1 presents results of studies in which data were collected at the student level, while Table 2 summarizes results from studies in which data were collected at the class level.[3]

The correlations and related statistical indices summarized in Table 1 represent within-course summaries of the relationship between grades and various measures of teaching effectiveness. The mean of the correlations reported between student grades and student evaluations of teaching in this table is approximately .21.

---

[3]The correlation coefficient measures the strength of the linear relationship between two variables. When the correlation coefficient is 1, an increase in the value of one variable exactly predicts the increase in the other. A correlation coefficient of 0 means that the variables are uncorrelated, or that an increase in one variable is not (linearly) associated with either an increase or decrease in the other. A value of -1 implies that an increase in one variable exactly predicts a *decrease* in the other. In most sociological applications, values above 0.2 or below -0.2 are considered substantively important.

This suggests a relatively strong relation between these two variables, particularly when one considers that account was not made for a large number of extraneous factors when the correlations in many of these studies were calculated. The natural effect of not accounting for such factors would be to diminish the estimates of the correlations reported.

The correlations summarized in Table 1 represent correlations between grades awarded to or expected by students within a class and the same student's evaluation of that class or teacher. Because all students within a class were presumably exposed to the same instructional techniques, differences in student evaluations of teaching that correlate with student grades might be assumed to result directly from a grading bias. Two mechanisms are commonly proposed to explain this bias.

The first, briefly mentioned above and referred to as the grade-leniency theory, was championed by many researchers in the 1960s and 1970s and embodies the notion that praise by an instructor for a student induces a favorable attitude by the student toward the instructor. Although this explanation for observed positive correlations between student grades and student evaluations of teaching is intuitively appealing to many professors, its popularity among educational researchers declined sharply with the advent of construct-validity research that began in the late 1970s and 1980s (e.g., [GG97]).

Disdain for the grade-leniency theory among educational researchers stems, at least in part, from its implications for the development of instruments to measure teaching processes. Broadly speaking, a fundamental problem encountered by many quantitative social scientists is the identification of measurable characteristics; the solution to this problem often involves the design of an effective survey instrument. For educational researchers attempting to measure teaching effectiveness, the most readily available subjects from whom information on teaching effectiveness can be gathered are students. Consequently, designing course-evaluation forms to measure various aspects of teaching effectiveness has become something of a cottage industry, and many of the most notable researchers in this area proposed their own instruments during the

1970s and 1980s. Suggestions that these instruments were biased by the effects of instructor grading policies were vigorously contested.

A second claim that the positive correlation between grades and teacher–course evaluations represents a biasing effect finds its source in attribution theory, as expounded by, for example, Gigliotti and Buchtel [GB90]. According to this hypothesis, individuals tend to be biased in their assignment of cause to events that affect them personally. Success is attributed to one's self, while failure is attributed to external sources. In the grading context, attribution theory predicts that students are more likely to attribute high grades and academic success to themselves, and to attribute lower than expected grades and academic failure to teachers and educational institutions. Thus, students who have difficulty mastering course material or are otherwise unsuccessful in a class will tend to blame the instructor for their failure, while better students will tend to attribute their success to hard work or intelligence.

Of course, positive correlations between student grades and student evaluations of teaching can also be explained by classroom mechanisms that do not represent a grading bias. For example, Remmers, Martin, and Elliot posited that

> ...the relationship between grades received in any individual class and the ratings given the instructor may not be direct, but the result of a third factor. This third factor may well be the level at which an instructor pitches his teaching. One whose teaching is "over the heads" of his weaker students may be held in high regard by his better students; in such a situation one might expect to find a positive correlation between grades and ratings, his instruction being at a level most appreciated by the students making the better grades. On the other hand, an instructor who directs his teaching to the weaker students and who is willing to spend an unlimited time in explaining relatively simple matters might be more "popular" with those students whose marks are relatively low; in such a situation one would probably find a negative correlation between grades and ratings [RME49, 25].

Other theories for the correlation between grades and student evaluations of teaching are based on the operation of unobserved,

intervening factors between grades and student ratings of instruction (e.g., [GG97]). For example, Marsh and others have proposed
that prior student interest plays an important role in creating a
positive correlation between grades and teacher–course evaluations
(e.g., [Mar83]). According to this conjecture, students with comparatively high levels of prior interest in a course will generally devote
more time to that course, and will also tend to be more appreciative
of the instructor's efforts. Such students will subsequently learn
more, and will thereby receive higher marks. In this case, prior
interest can be regarded as a biasing effect on student evaluations
of teaching, but the assignment of higher grades is not.

Similarly, general student motivation might be considered as
an intervening factor that leads both to greater student learning and higher grades. This explanation was invoked to explain
results reported in the studies of Howard and Maxwell [HM80]
and Marsh [Mar84]. As in the case of prior subject-matter interest,
differences in motivation of students within a class are not considered to be a grade bias if more highly motivated students tend to
learn more, to receive higher grades, and to value the efforts of the
instructor more highly.

A final intervening factor that has been proposed to explain
the positive correlation of assigned student grades and student
evaluations of teaching is good teaching. Ideally, this is precisely the
factor that student evaluations of teaching are designed to measure.
Correlations that result from the causative relationship between
good teaching and student learning—assignment of higher grades—
is desirable. *Unfortunately, this factor, and the teacher-effectiveness
theory it engenders, cannot explain the positive correlations reported in
Table 1 because in these studies all students whose grades were used
to compute the within-class correlations presumably received the same
instruction.*

Ironically, prior subject interest, higher levels of student motivation, and other intervening factors (aside from teacher effectiveness) that might explain the correlation between student grades
and student evaluations of teaching are often also used to argue
against adjustments of teacher–course evaluations to account for
the positive correlations between grades and course evaluations. In

an example that typifies this reasoning, Marsh [Mar83] presented a path analytic model in which prior subject interest, the effect of prior subject interest through workload/difficulty, reason for taking a course, and the effect of reason for taking a course through workload/difficulty were all argued to be causes of the observed correlation between student grades and teacher–course evaluation factors, as measured, of course, by Marsh's SEEQ form. Based on this path analysis, Marsh concluded that the observed positive correlation between teacher–course evaluations and grades was not due to a grading bias. In a later article, he went on to characterize the search for biasing factors in student evaluations of teaching as a "witch hunt" ([Mar84, 730]). Marsh's reasoning, repeated in sundry variations in many of the studies cited in Table 1 for which non-statistically significant partial correlations were reported, is that the absence of a grading bias makes teacher–course evaluations unbiased.

The fault in this reasoning is that any intervening factor that causes a correlation between grades and course evaluations also represents a bias. By definition, such intervening factors are beyond the control of the instructor, and though they may represent valid predictors of student learning, their uneven distribution in students across classes certainly results in a bias in the assessment of teacher performance. An instructor faced with a class of unmotivated or uninterested students is as unlikely to obtain favorable teaching evaluations as is an instructor who alienates a class by assigning unusually stringent grades. Indeed, the teacher in the former circumstance has less control over his evaluations than does the latter.

As stated above, the teacher-effectiveness theory cannot explain within-class correlations between student evaluations of teaching and assigned or expected student grades. All alternative explanations of this correlation—the grade-leniency theory, attribution theory, theories that rely on intervening factors—as well as modifications of the teacher-effectiveness theory as proposed by, for example, Remmers, Martin, and Elliot [RME49], predict a change in student evaluations of teaching concordant with changes in assigned or expected student grades. Thus, adjustments to teacher–course evaluation scores to account for these within-class correlations are clearly appropriate if these scores are to be used to compare the performance of teachers whose assigned grade distributions differ.

Similar comments apply also to between-class correlational studies, which are summarized in Table 2. There, the unit of analysis is the class. Correlations and other indices of association between grades and student evaluations of teaching in this table were computed by examining the relationship between the mean grades assigned across several classes and the corresponding means of responses on teacher–course evaluation forms. Approximately one-half of the studies summarized in this table report correlations between global measures of teaching effectiveness and mean course grades; the remainder report correlations between mean course grades and specific dimensions of teaching performance. The mean of the correlations reported in studies in this table is approximately 0.31. As in Table 1, this mean correlation suggests a strong relation between grades and student evaluations of teaching.

In principle, all theories proposed to explain the positive within-class correlations observed in Table 1 can be extended to account for the between-class correlations reported in Table 2, although the plausibility of each must be reassessed to account for the change in unit of measurement, i.e., from students to classes.

The teacher-effectiveness theory provides a viable explanation for the correlations in Table 2, since positive correlations between mean grades and teacher–course evaluations could result if better teaching led to both higher grades and higher evaluations of teaching. Note, however, that this effect could be masked or even inverted if, for example, in a given study, better teachers graded more stringently than their less able counterparts.

Like that of the teacher-effectiveness theory, the feasibility of the grade-leniency theory as an explanation for the correlation between grades and student evaluations of teaching also increases at the class level of analysis. Within classes, the operation of the grade-leniency theory implicitly requires that students believe that their instructors are differentially lenient in the assignment of their grades in relation to other students in the same class. In other words, for this theory to operate within classes, students must believe that instructors assign grades in a highly subjective manner, even one showing favoritism.

TABLE 2

Class-Level Studies of Grades and SETs. Values in this table summarize published correlations between expected or received classroom mean grades and mean classroom student evaluations of teaching. Studies that reported significant (highly significant) t or F statistics have $*$ ($**$) inserted into the column marked "correlation." Statistically significant ($p < 0.05$) and highly significant ($p < 0.01$) correlations are indicated by $*$ and $**$.

Author (date)	Correlation	Comments
Aleamoni & Spencer [AS73]	$< .30*$	Small, but significant, correlations of items on a factor-based instrument, CEQ, with actual grade received for 2,784 classes containing 100,000 students taught at the University of Illinois
Anikeef [Ani53]	.43, 73**	Spearman's ρ based on ranks of faculty grading leniency (as determined by mean grade assigned) and aggregated teacher/course rating scores
Bausell & Magoon [BM72]	**	Used one-way analysis of variance to assess effects of expected grade and "discrepant" grade on 29 course evaluation items; both expected and discrepant grade were highly significant predictors on 26 of 29 items
Bendig [Ben53]	.14, .28**	Between-section correlations of actual grades and instructor and course ratings, respectively
Brandenburg, Slinde, & Batiste [BSB77]	.42**	Correlation between expected grade and overall course rating

TABLE 2 *(continued)*

Author (date)	Correlation	Comments
Brown [Bro76]	.35*	Correlation of actual grade with mean course rating; also discusses regression analysis in which course grade enters equation last, but increases multiple R^2 value from .25 to .39, even after account of numerous course–teacher attribute variables
Frey [Fre73]	.69**	Correlation of adjusted final exam grade for class (adjusted for SAT) and average course evaluation score
Frey [Fre76]	.75	Correlation between average course–teacher evaluation based on six dimensions of teaching and final exam performance
Frey, Leonard, & Beatty [FLB75]	.54	Average correlation, across multiple section courses taught at three universities, of final exam performance and six course attribute variables; "workload" correlated negatively with final exam performance and was excluded from the average
Gessner [Ges73]	.62*, .74*	Correlation with course content and presentation and mean class performance on national exam
Greenwald & Gillmore [GG97]	.38* to .50*	Standardized path coefficients (on scale of correlation) relating expected grade to course evaluation

TABLE 2 *(continued)*

Author (date)	Correlation	Comments
Howard & Maxwell [HM80]	.09 to .27	Range of partial correlations reported for regressions of student satisfaction variables on expected grade
Kennedy [Ken75]	**	F-tests of the effects of expected, actual, and difference between actual and expected grades were conducted to assess the impact of these variables on three dimensions of teaching effectiveness; significant positive correlations were noted for all effects; actual grade produced strongest and most consistent correlation with positive course evaluation across all three dimensions
Kooker [Koo68]	**	F-tests of the effect of actual grade on seven teaching dimensions for both upperclass and first-year students; with the exceptions of "textbook" and "presentation" among upperclass students, which were not significant, highly significant positive correlations between grade received and teaching effectiveness ratings were noted in all other cases
Krautmann & Sander [KS99]	**	Highly significant and marginally significant ($p < .10$) positive coefficients reported in ordinary and two-stage least squares regressions of mean

TABLE 2 *(continued)*

Author (date)	Correlation	Comments
		overall teacher effectiveness on mean expected grade
Landrum [Lan99]	.19**, .28**	Correlations of expected grade and overall rating of course and instructor, respectively
Marsh [Mar83]	.21	Average correlation reported across eight dimensions of teaching effectiveness, averaged across academic disciplines and teaching dimensions; range of reported correlations extended from -0.02 to .42
Mirus [Mir73]	$.49_1$**	Partial correlation of regression of professor rating versus expected grade in course
Nelson & Lynch [NL84]	.27*	Correlation of mean expected grade and overall measure of course quality; correlation of mean expected grade and overall instructor was .15, but was not statistically significant; authors also report significant, positive regression for mean expected grade on course quality in three-equation simultaneous regression, and insignificant, positive coefficient for mean expected grade on instructor evaluation
Pohlmann [Poh75]	.42**	Average correlation of expected grade and five dimensions of teaching effectiveness

TABLE 2 *(continued)*

Author (date)	Correlation	Comments
Rayder [Ray68]	−.01, .14, .18	Correlation between previous grade awarded by instructor and three teaching traits (responsible behavior, friendly behavior, and enthusiasm)
Remmers, Martin, & Elliot [RME49]	.27*	Average reported correlation between differences in expected and received grades and course rating
Riley, Ryan, & Lifshitz [RRL50]		Analysis of student ratings of instructors on ten dimensions of teaching effectiveness as a function of academic standing; in nine of ten dimensions examined, authors conclude that evaluations increase with academic performance of student
Rodin & Rodin [RR72]	−.75*	Partial correlation of teaching assistant's ratings and mean course grade adjusted for initial ability level (determined from previous course)
Rubenstein & Mitchell [RM70]	.12	Correlation of actual grade and instructor evaluations
Savage [Sav45]	.15**	Correlation of course ratings and actual student grades, aggregated across classes.
Scheurich, Graham, & Drolette [SGD83]		No significant correlation reported between mean expected grade and overall instructor evaluation in multiple regression framework; a significant, simple

TABLE 2 *(continued)*

Author (date)	Correlation	Comments
		correlation of .48 between these variables is reported.
Shapiro [Sha90]	.28**	Correlation between mean expected grade and overall evaluation of course; positive, statistically significant coefficients also reported for two multiple regression models
Stewart & Malpass [SM66]	**	One-way analysis of variance of the effect of expected course grade on seven teaching dimensions, including severity of grading practice; highly significant findings reported within each teaching dimension; higher expected grades were associated with higher evaluations.
Sullivan & Skanes [SS74]	.39*	Mean correlation of common final scores and overall instructor evaluation in 124 sections of first-year science courses
Weaver [Wea60]	**	One-way analysis of variance of the effect of expected grade on "teaching technique" and "instructor personality" and "total" ratings; highly significant effect reported for teaching technique and total ratings, insignificant effect reported for personality traits variable

In contrast, grade-leniency can operate at the class level of analysis whenever students recognize that different instructors employ different grading policies. If students rate highly those instructors whom they feel grade less stringently, positive correlations between mean class grades and student ratings of teaching would occur.

The extension of attribution theory and theories based on intervening factors, like prior student interest and student motivation, translate more directly across the two types of studies, although aggregation to the class level affects each theory differently.

The probable effects of prior student interest and, to a lesser extent, general student motivation depend on the level and types of classes included in a study. For example, at most universities first- and second-year students are generally assigned to introductory-level courses in an almost random way. Because of this randomization, major differences in average prior student interest or general student motivation between classes taught in the same department are unlikely to be important at the class level of analysis. As a result, such intervening factors are probably not responsible for the correlations between grades and teacher–course evaluations observed in lower-level, introductory courses taught within the same department.

Prior student interest and student motivation provide more reasonable explanations for positive correlations between student grades and student evaluations of teaching in upper-level courses because students are generally permitted to select such courses themselves. Thus, average student interest and motivation levels can vary significantly between courses. Courses with high levels of prior student interest tend to produce higher grades and higher course evaluations, while courses with less-motivated and less-interested students tend to generate lower grades and lower course evaluations. Undoubtedly, the effects of prior student interest and student motivation are at least partially responsible for the higher grade distributions often reported in upper-level courses.

The interpretation and operation of attribution theory remains essentially the same at both the class and student levels of analysis. According to this theory, students in classes with relatively low grade distributions will tend to assign blame for their low marks to instructors. This leads to lower than average teacher–course evaluations, which in turn result in a positive correlation between grades and student evaluations of teaching.

It is noteworthy that the correlations reported in Tables 1 and 2 tend to have similar signs and magnitudes. An overwhelming majority of studies in both tables report either statistically significant positive correlations or statistically insignificant correlations between grades and student evaluations of teaching. Point estimates of the statistically insignificant correlations also tended to be positive. The one notable exception to this trend is the study of Rodin and Rodin [RR72], in which a highly statistically significant partial correlation of −.75 was reported.

Because the Rodin and Rodin finding runs contrary to all prevalent theories that predict correlations between grades and teaching effectiveness, it is not surprising that this study was the source of substantial controversy when it first appeared in the journal *Science* in 1972. Its deviant behavior also warrants some discussion here.

The basic design of the Rodins' study was as follows. Students in a large introductory physics course, taught by a single professor (Burton Rodin), were assigned to 12 recitation sections taught by 11 teaching assistants. Each recitation section met twice weekly. One session was devoted to answering questions about lectures, while the other was used to administer quizzes and to review solutions to previous quizzes. Each quiz was designed to measure students' mastery of a specific task, or paradigm problem, and quizzes were graded as being either completely correct or incorrect. Students who failed a quiz on a particular topic could retake quizzes on the same paradigm problem up to six times. Final grades in the course were determined by the number of topics mastered, or, equivalently, the number of quizzes successfully passed. The number of failed quizzes had no bearing on final course grade. The performance of each teaching assistant was judged by the mean response of his students to the question "What grade would you assign to his total teaching performance?" Responses to this question were collected at the end of the quarter at the common lecture session.

After accounting for student abilities using grades received in a previous quarter, Rodin and Rodin reported a partial correlation between average teaching assistant rating and grades in the recitation sections of $-.746$; the simple correlation between these variables was $-.754$. In effect, teaching assistants who taught better—or at least who had students that received higher grades—received lower ratings, causing the Rodins to conclude that "good teaching is not validly measured by student evaluations in their current form" [RR72, 1166].

Several anomalies in the Rodin and Rodin study were subsequently used to explain the Rodins' surprising conclusion. Frey [Fre73] noted that the evaluation of teaching assistants was based on a single global survey item, and contended that different results may have been obtained had the wording of this item been different. Frey also questioned the reliability of the data, noting that the variability between the two classes taught by the same teaching assistant signaled a high degree of unreliability. Sullivan and Skanes [SS74], Frey [FLB75], and Gessner [Ges73] questioned the validity of measuring the performance of teaching assistants who were not primarily responsible for the organization of the course, for determining the paradigm problems tested, or for setting the quizzes. Another explanation for the Rodins' unusual finding might simply be that better students were annoyed by the requirement to review previous quizzes and material before taking the current quiz, and so were unappreciative of the assistance offered by the teaching assistants. This is basically the explanation proposed by Rodin, who, citing Remmers, Martin, and Elliot [RME49], suggested that the teaching assistants had pitched their instruction to lower-performing students ([Rod75]; see also [Doy75]).

One lesson that can be learned (or at least reinforced) from the furor caused by the Rodins' study is that all observational studies are, in some sense, flawed. Many of the criticisms leveled against that study apply equally well to many of the other studies cited in Tables 1 and 2. For example, while it is true that the Rodins' study relied on the analysis of a single global evaluation item rather than analyses of well defined factors measuring distinct teaching dimensions, so, too, did approximately one-half of the other studies in these tables. It is also true that the teaching assistants in the Rodins' study were not intimately involved in establishing the syllabus for the course or in determining the questions used to test student achievement. However, similar criticisms can also be directed at almost all multisection validity studies, which are discussed in Chapter 5 and are widely touted by educational psychologists as providing the best method for establishing construct validity of survey instruments. In short, the flaws exposed in Rodin and Rodin are also common in many other correlational studies.

More generally, correlational studies are, by their nature, unable to account for numerous unrecorded environmental variables that confound the measurement of intended relationships. In the context of educational assessment, numerous environmental factors prevent the establishment of causal links between grades and student evaluations of teaching. Aside from the confounding factors already discussed, important environmental variables not considered in any or most of these studies include the following:

Grades received concurrently by students in other classes

According to the grade-leniency and grade-attribution theories, students will rate most highly those instructors from whom they receive the highest grades. However, none of the studies summarized in Table 1 account for other grades received by students during the semester(s) in which the studies were conducted, and few even consider student GPAs. Clearly, a student who expects a grade of C in a course will view such a grade more positively if he expects to receive D's in all his other courses, even if the mean class grades in all of his courses are comparatively high.

Differences in grade distributions between classes

Within-class correlations between grades and student evaluations of teaching are attenuated when the grade distribution of a class deviates from the norm. For example, all students within a class in which the instructor assigns harsh grades might be disappointed with their grades, whereas all students in classes with a high grade distribution might be satisfied with their grades. If the relationship between grades and teacher–course evaluations saturates above or below some threshold, so that there is no additional effect of a student receiving say, a C− rather than a C, estimated within-class correlations will disappear for classes with mean grades at the extremes. Conversely, there is some evidence to suggest that students occasionally resent abnormally high grade distributions when such distributions are not justified by correspondingly high workloads or course content [Abr85].

Differences in grade distributions by field of study

Students may calibrate their expectation and interpretation of assigned grades according to perceived differences in grading norms across different academic disciplines. For example, students might interpret an A in a physics course differently than an A in a language skills course if they believe that an A is harder to obtain in physics.

The effect of each of the confounding factors described above is to *attenuate* the estimates of the correlation between student grades and student evaluations of teaching. Indeed, with so many pitfalls possible, the relative consistency of reported correlations in Tables 1 and 2 is somewhat surprising (see Note). Unfortunately, observational studies provide little insight into the actual cause of the correlations reported in Tables 1 and 2, for reasons discussed above. With the exception of the teacher-effectiveness theory, all theories provide plausible explanations of the correlations in Table 1; the same holds true without exception for correlations reported in Table 2.

EXPERIMENTAL STUDIES

U nderstanding the mechanism that produces the correlation between grades and student evaluations of teaching is important because the nature of this mechanism determines the nature of reforms that must be implemented to correct it. If correlations between grades and student evaluations of teaching result from differences in teacher effectiveness, then corrections to teacher–course evaluations for differences in assigned grades are both unnecessary and inappropriate. If, on the other hand, positive correlations between student grades and student evaluations of teaching are explained by student attribution or grade-leniency

NOTE

Despite the general consistency of findings in Tables 1 and 2, educational psychologists are loath to acknowledge the impact of grades or other intervening factors on student evaluations of teaching. This aversion manifests itself throughout the educational research literature, and articles in which these effects are discounted are almost as plentiful as studies in which these correlations are measured. The following discussion in two particularly influential articles illustrates the general tone of this literature.

In their seminal article on the reliability and validity of student ratings of teaching, Costin, Greenough, and Menges summarize two earlier studies on the relation between grades and student evaluations of teaching as follows: "Spencer (1968) found that correlations between grades and 'overall value of course' seldom exceeded .30. Similar low correlations, ranging from .23 to .32 were reported by Caffrey (1969) for six dimensions of student ratings: overall value of course, skill in instruction, use of class time, anticipation of student difficulties, friendliness of students in class, and clearness of explanation." They conclude from these findings that "the fact that the positive correlations which were obtained between student ratings and grades were typically low weakens this claim as a serious argument against the validity of student ratings" [CGM71, 519].

A casual reader of this article might conclude that correlations in the range of .3 are thus not to be regarded as a cause for serious concern. However, context can be important, as witnessed by their comments several paragraphs later when they turn their attention to establishing the validity of student ratings vis-à-vis student gains in knowledge. "Although measuring student gains as an index of 'good teaching' is fraught with practical and technical difficulties, and although it would be difficult to apply in many teaching situations, the approach warrants serious consideration. In one of the few studies illustrating such an approach, Morsh, Burgess, and Smith (1956) found that students in an aircraft mechanics course at an Air Force installation made gains in information and in practical 'job sample' performance which were significantly correlated with their overall ratings of the course (r=.32 for information and .39 for practical performance). Slightly higher correlations were found for students' rating of teaching ability (.41 for each criterion gain)" [CGM71, 520].

Bearing in mind that the Morsh, Burgess, and Smith study involved over 3,000 Air Force recruits assigned nearly at random to small training classes as part of their advanced training program, and that learning gains in this study were extremely well defined by end-of-course objectives, and that all instructors were aware of testing criteria, it is hard to imagine a better setting for measuring the effects of teaching on student learning. Still, the largest correlation found between student evaluations of a given teaching dimension and student performance was .41, not quite twice the value of the correlations between grades and student evaluations that the authors previously dismissed.

As a second example in an equally influential paper, Marsh [Mar82] reported that 12–14% of the variance in student evaluations of teaching can be explained by background variables like prior student interest and student grades. This figure represents the partial *squared* correlation, which implies a raw correlation in the range of .35–.37. However, the correlations cited by Marsh [Mar84] in summarizing a meta-analysis of validity results performed by Cohen [Coh81] range from .5 for instructor skill to .43 for overall instructor (and near zero with course difficulty). Marsh used such correlations to argue in favor of the validity of the multitrait approach to teaching evaluation. Once again, grades and other intervening variables accounted for nearly as much variation in student evaluations of teaching as did more salient measures of teacher-effectiveness, yet the effects of grades and other intervening factors were not regarded by the researchers as important variables in the interpretation of student evaluations of teaching.

effects, corrections to teacher–course evaluations are needed to avoid the repercussions predicted by Winsor [Win77] and others. Causation by intervening variables is also troubling; if intervening factors like prior student interest or student motivation are responsible for these correlations, then adjustments to student evaluations of teaching to account for these effects are needed before student surveys are used in faculty promotion, tenure, and salary reviews.

Numerous experiments have now been conducted to untangle the causal relationship between student grades and student evaluations of teaching. Although many of these experiments were conducted in rather artificial, laboratory-like settings, several others involved direct manipulation of student grades in actual college environments.

In one of the first such published experiments, Holmes [Hol72] studied the relationships between disconfirmed grade expectancies, course grades, and instructor evaluations through an experimental design that manipulated student grade expectations. In this design, each student was told that his or her final course grade would be determined by scores received on four multiple-choice tests administered during the semester. After each of the first three tests, students were given their actual numerical score and a suggested grade distribution for these scores. On the last examination, students were asked what grade they expected to receive in the course after being told that their response to this question would have no effect on their final course grade.

After grading the final test, Holmes selected subjects whose actual and expected grades were identical, with both being either an A or B. Students in each grade group were then randomly assigned to be either experimental subjects or controls. The numerical score reported to the experimental subjects for the final exam was artificially lowered so as to reduce their course grade by one mark. Subsequently, students were informed of their grade on the final test and were informed of the mapping that would be used to convert their numerical average to a course letter grade. Students were not informed that an experiment was being conducted. In addition, the mapping from numerical score to grade was consistent with the earlier grade distributions. Thus, experimental subjects could not attribute their lower than expected scores to a change in instructor grading criteria.

At the end of the last class, students were asked to complete an end-of-course teaching evaluation form. Based on data collected from this form, a two-way analysis of variance was performed to determine the impact of grade (A or B) and disconfirmed grade expectancy on each of the evaluation items.

The results of the statistical analysis of these data suggested that grade was not a significant factor on any of the 19 items appearing on the course evaluation form, although only students whose actual grades were A or B were incorporated into the experiment. The number of students receiving a grade of C or worse in this class was apparently small, and the author states that these students' grades could not be "credibly manipulated" to the D level ([Hol72, 131]). Whether this suggests that the overall grade distribution in this class was high in comparison to other classes in which students were simultaneously enrolled cannot be determined from the data provided.

The effect of disconfirmed grade expectancy (i.e., being told that one's received grade was lower than that actually earned) was statistically significant at the .05 level of significance on 5 of the 19 evaluation items, including questions regarding the preparedness of the instructor, whether the instructor had sufficient information to evaluate their achievement, whether the instructor was intellectually stimulating, whether the course had met their expectations, and whether examination questions were clear. In addition, the effect of disconfirmed grade expectations exhibited a strong trend ($p < .10$) on five other items measuring coherency of presentation, promptness in returning assignments and tests, motivating students to work on course material, and holding student attention. On all ten items, the direction of the effect was as expected; students whose grades were manipulated downward gave lower ratings than students whose grades were not altered.

Because of the manner in which grades were manipulated during this experiment, it is difficult to argue that teaching effectiveness or other intervening factors, like prior student interest or general student motivation, could have caused the observed differences in the ratings between experimental subjects and controls. Grade leniency and grade attribution theories provide viable explanations for these differences, as does the closely related theory forwarded by Holmes; namely, that disconfirmed grade expectancies cause students to deprecate instructors' teaching performance.

Holmes's study was extended in a later experiment performed by Worthington and Wong [WW79]. In the later study, students were divided into three groups based on grades received on a practice test. Within each of these groups, students were randomly assigned a grade of either "good," "satisfactory," or "poor." After receiving their practice grades, which they were told would predict well their final course grade, the students were asked to complete a teacher–course evaluation form that was labeled surreptitiously with their assigned grade and earned grade categories. The authors present several analyses of collected course ratings by treating different combinations of assigned and earned grades as factors. The most striking analyses involve the comparison of those students randomly assigned a grade of satisfactory versus those assigned a grade of poor. In this comparison, students who received the satisfactory grade rated the instructor or course more highly on 17 of the 18 evaluation items than did those students who received a grade of poor.

In a third study of this type, Blunt manipulated grades of students in an introductory psychology course taught at a community college. Blunt also introduced anonymity as a second treatment effect by requiring approximately one-half of the students in the course to sign their evaluation forms. According to this study's experimental design, students were provided with their current course grade—based on a midterm and three course assignments prior to the final examination—and were then asked to complete a short form of the Purdue Rating Scale for Instruction. Students were randomly assigned into four grade treatment categories depending on whether they were given an inflated grade, a deflated grade, their actual grade, or no assigned grade. The results from this experiment were analyzed using an analysis of variance procedure. Two of the 10 items reported on the evaluation form were significantly affected by the grade treatment. The instructor was "rated as more clear, definite, and forceful when presenting subject matter by the inflated- and true-grade groups than by the deflated- and no-assigned-grade groups, [$F(3, 90) = 2.90, p < .04$]," and the degree to which the instructor stimulated intellectual curiosity was "highest for the inflated-grade group and lowest for the deflated-grade group [$F(3, 90) = 4.82, p < .01$]" [Blu91, 49].

In yet another grade-manipulation experiment, Vasta and Sarmiento [VS79] report results from a study in which the grade distributions of two sections of an introductory psychology course were manipulated. In their experiment, the grade curves used to map scores on multiple-choice tests to letter grades were altered so that one section received a substantially higher average grade than the other. Each section was jointly taught by both authors, and Vasta was unaware of which section had been assigned the more lenient grade distribution. To eliminate the possibility that treatment differences discovered at the end of the study might also be be attributable to differences in mean student characteristics between sections, the authors conducted numerous statistical comparisons of the students in the two sections and found no statistically significant differences in either the study habits, grades, or demographic composition of students in the two classes.

After informing students of their test grades at the end of the final unit in the course (but before the final exam), students in each section were asked to complete an anonymous course-instructor evaluation form consisting of 50 items designed to measure various properties of the course and instructors (18 course-related items and 16 instructor items that were completed for both instructors). Of these items, seven showed statistically significant differences in mean response between the two sections ($p < 0.05$), and in each case the section that was graded more leniently responded in a more favorable way. These items concerned ratings of reading material, lengths of assignments, balance between lectures and discussion, stimulation of interest, the extent to which the instructor required independent thinking, rapport with class, and anticipated performance on exams. Section differences on an additional four items approached significance ($p < .10$), and on three of these items the group with the more lenient grade distribution again provided more favorable ratings. These items included questions that probed instructors' overall teaching ability, preparation for class, and whether or not the student would recommend the class to a friend. Interestingly, the adequacy of exam coverage was rated contrary to expectations. Finally, by examining only the sign of the difference between the mean response of each section, 16 of 18 course items and 24 of the 32 instructor items were rated more highly by the leniently graded class ($p < 0.01$ in both cases).

Vasta and Sarmiento are careful to emphasize that the grade manipulations used in their study were comparatively weak, since on any given exam, only about 60% of the students were affected by the manipulation, and the same students were not necessarily affected on subsequent or previous exams. In addition, the magnitude of the manipulation was comparatively small, resulting in at most a one letter grade difference in any student's grade. They conclude that the direct, causative effects of grades on student evaluations of instruction are "potentially quite powerful" [VS79, 210].

A secondary, but nonetheless noteworthy, finding of the Vasta and Sarmiento study was that the grade manipulations did not appear to affect the study habits of students in sections in which grades were manipulated downward. Several explanations are offered to explain the lack of an effect of manipulated grades on student study patterns, including the possibility that some students might work harder to overcome the deficit, while others might become "dispirited and do even less" work in the future. In any case, the lack of an association between grades and study habits suggests that the motivating effect of grades on student effort is not fully understood [VS79, 211].

Chacko [Cha83] employed methodology similar to that of Vasta and Sarmiento to measure the effects of grades on student evaluations of teaching. In Chacko's experiment, two sections of the same course, taught by the same instructor, were given nearly identical instruction and exactly the same midterm examinations. However, the grades received by students on the exam in one section were norm-referenced so as to produce a higher grade distribution. The severity of this manipulation ensured that no students in the norm-referenced section received a grade lower than a B, whereas the exam was so difficult that no student in the uncurved section received an A. The numerical grades in the two sections were almost identical (means of 48.5 and 48.8, respectively), suggesting that there were probably no large differences between sections in either the effectiveness of teaching or in student background variables that might have affected student performance.

To assess the impact of grades on student evaluations of teaching, a teacher–course evaluation form was administered to students in this experiment on two occasions: one week before the midterm, and again one week after the midterm grades were announced.

The particular course-evaluation form used in this study included items concerning preparation of instructor, course organization, presentation, explanation, knowledge of subject matter, intellectual motivation, self-reliance and confidence, sense of proportion and humor, availability, and attitude toward students. On none of these items were significant differences found between sections on the pretest survey. However, significant ($p < 0.05$) or highly significant ($p < 0.01$) differences between sections were noted in 7 out of 10 of these items on the posttest questionnaires, and in each case the more stringently graded section reported less favorable ratings. On the remaining three items (organization, presentation, and availability), observed differences in section means also suggested less favorable ratings by the more stringently graded section, although statistically the differences were not significantly different from zero. All items were rated on a seven point Likert-type scale; the average mean difference in the section ratings, averaged over the 10 evaluation items, was approximately .6 units per item.

A potential flaw of the Chacko and Vasta–Sarmiento studies that should be noted involves the possibility that some students may have discovered that different grading policies were used in the control and treatment groups. If this occurred, observed differences in student evaluations of teaching might have resulted in part from a perception of unfairness by students enrolled in the more stringently graded sections. However, neither article mentions student complaints in this regard, and the effects, if any, of contamination from a small number of students who may have discovered the deception would probably have been small.

SUMMARY

Concern over the potential for student grades to bias student evaluations of teaching has a long history, dating back at least to studies conducted by Remmers in the late 1920s [Rem28]. Since that time, scores of observational studies have been performed to investigate the plausibility of these concerns, and an overwhelming majority of these studies have identified positive correlations between student grades and student evaluations of teaching. On average, the magnitude of these correlations has typically ranged from 0.2 to 0.3, but has varied substantially from study to study. Still, no consensus has yet been reached by educational researchers regarding the cause of this correlation.

Much of the variation observed in the correlation coefficients reported across studies can undoubtedly be attributed to variation in environmental factors. For example, systematic differences in prior student interest, student motivation, level of course, class size, differences in grading norms between academic disciplines, demographic variations in student populations, and differences between student grade expectations all potentially affect either student grades or student evaluations of teaching. With so many possible intervening factors at play in these studies, and so few controls imposed to account for their effects, the consistency of the positive correlations that has been reported is, in a certain sense, unexpected and likely portends a significant causal effect of grades on student evaluations of teaching.

Unfortunately, observational studies are ill suited for determining the source of the generally reported positive correlation between student grades and student evaluations of teaching. Several theories have been proposed to account for this correlation, including the teacher-effectiveness theory, theories based on the intervening variables, the grade attribution theory, and the grade-leniency theory, but it is difficult to distinguish between the operation of each theory based only on observed correlations. This makes the interpretation of teacher–course evaluation forms difficult, since the validity of the information provided relies heavily on the causal nature of this relationship.

In addition to the observational studies summarized in Tables 1 and 2, several grade-manipulation experiments designed to more directly measure the effects of grading practices on student ratings of instruction were discussed. In all of these experiments, decreases in expected grades, received grades, or differences between received and expected grades were found to adversely affect student evaluations of teaching.

In contrast to the correlations reported in observational studies, it is difficult to ascribe the effect of grades on student evaluations of teaching reported in the grade-manipulation experiments to any cause other than differences in instructor grading practices. With the exception of grading policies, course attributes of the treatment and control groups were essentially identical in each of these experiments. Furthermore, grades were found to have a statistically significant and substantively important effect on some subset of the items contained on each of the course evaluation forms employed. Thus, these experiments imply that grading practices cause changes to student teacher–course ratings that cannot be explained either by teacher effectiveness or through the operation of intervening factors like prior student interest or motivation.

DUET Analysis of Grades and SETs

In this chapter, DUET experimental data are used to clarify two issues that have clouded the debate over the use of student evaluations of teaching in administrative reviews of faculty since their introduction to widespread use in the early 1970s. The first issue concerns the extent to which grades reflect a "biasing" influence on student evaluations of teaching. By examining patterns of student response to the DUET survey as a function of student grades, prior interest in course subject matter, class mean grades, and the consensus ratings of items as estimated from other students' responses to the survey, the correlation between student grades and student evaluations of teaching is shown to result primarily from grade attribution, the process by which students associate their success in a course with the grade they receive.

The second issue involves the magnitude of the bias that grades have on student evaluations of teaching. Using responses to the DUET survey collected from the same students both before and after they received their final course grades, the influence of grades on student evaluations of teaching is isolated from external factors that have confounded earlier studies. Analyses based on these data demonstrate that the effects of grades on teacher–course evaluations are both substantively and statistically important, and suggest that instructors can often double their odds of receiving high evaluations from students simply by awarding A's rather than B's or C's.

DESPITE NEARLY EIGHTY YEARS OF EMPIRICAL research probing the relationship between student grades and student evaluations of teaching, no consensus has yet emerged regarding the extent to which student grades exert a biasing influence on student evaluations of teaching. This lack of consensus may be attributed to a variety of factors, most of which involve challenges to the methodology employed in past studies. As demonstrated in the previous chapter, observational studies have firmly established the existence of a positive correlation between grades and student evaluations of teaching, yet they have proven inappropriate for establishing a causal link between these variables. Experimental evidence is less ambiguous, but is also less abundant and has often been collected under less than ideal circumstances. In this chapter, new insights into the connections between student grades and student evaluations of teaching are exposed using data collected during the DUET experiment.

The DUET data provide a mechanism for examining two issues critical for understanding the correlation between grades and student evaluations of teaching. The first involves the cause of this correlation. Because the DUET response data were linked to the students who supplied it, it is possible to examine the way student responses to the survey varied according to the characteristics of participating students, their success in their courses, and the attributes of these courses as reported by other students. In particular, it was possible to evaluate the relative importance of student grades, prior student interest, instructor grading practices, and actual teaching behaviors in determining how students responded to survey items. By so doing, the extent to which expected or received grades caused changes to student evaluations of teaching could be determined.

Second, the magnitude of the causal effect of grades on teacher–course evaluations was estimated using the DUET data. In many courses, evaluation data were collected twice from the same students, once before and once after they had received their final course grades. By examining the way student responses changed according to differences in the grades they expected to receive in the fall and the grades they ultimately did receive in the spring, the influence that grades had on student evaluations of teaching was isolated from external variables—like student aptitude, interest, and motivation—that have confounded the evaluation of grade effects in past studies.

To resolve these issues, two approaches to analyzing the DUET data are discussed below. In the first, standard linear regression analyses are employed to qualitatively assess the relative importance of course quality, student grades, prior student interest, and mean course grades on student evaluations of teaching. The primary goal of this analysis was to assess the validity of various grading theories in explaining the relation between grades and student evaluations of teaching. Secondly, ordinal regression models were used to estimate the magnitude of the effect of student grades on student evaluations of teaching by examining changes in student responses that accompanied changes in their expected and received grades.

INTERVENING VARIABLES AND STUDENT EVALUATIONS OF TEACHING

Several explanations are commonly offered to explain the generally observed positive correlation between student grades and student evaluations of teaching. These explanations include the grade-leniency theory, the grade-attribution theory, the teacher-effectiveness theory, and explanations based on intervening variables like prior student interest, motivation, and aptitude. In this section, the utility of each of these theories in predicting observed patterns of student responses to DUET items is examined by regressing survey responses on variables relevant to the operation of each theory.

The methodology applied to this problem, regression analysis, is a statistical technique used to quantify the extent to which changes in one or more explanatory variables foreshadow changes in a response variable. A simple example of a regression analysis might involve, say, the prediction of house prices based on floor space, measured in square feet. For a particular neighborhood, a "best-fitting" regression equation might take the form

$$\text{house price} = 125 \times \text{square footage} + \text{error}.$$

The interpretation of this equation is that, on average, a 100 square foot increase in floor space is associated with a \$12,500 increase in the cost of a home. The error term accounts for the fact that home prices within a neighborhood are determined by more than simply the square footage.

More generally, if we were to let H_i denote the price of the ith house being considered, S_i its area in square feet, and e_i the error that results in predicting a house's price based only on its square footage, this same relationship can be written more compactly as

$$H_i = \beta S_i + e_i.$$

In this expression, the regression coefficient β represents the average rate at which home prices in the neighborhood increase with each one square foot increase in their area. In this case, β is equal to \$125 per square foot.

A similar approach can be taken to assess the relative impact of student grades, classroom mean grades, and student's prior interest on student responses to items on the DUET survey. In this context, the response variable of house price is replaced with the standardized responses of students to items on the DUET survey. In place of square footage, the explanatory variables are standardized values of students' expected or received grades, their prior interest in course material,[1] the mean grade awarded in the course, and estimates of the "true" item response, as estimated from the responses of other students. As in the case of house prices, this relationship may be expressed more compactly by introducing variables to represent each of the relevant quantities. Doing so yields an equation of the following form:

$$Y_{i,j} = \beta_G G_{i,j} + \beta_R R_j + \beta_P P_{i,j} + \beta_A A_{i,j} + e_{i,j}. \tag{4.1}$$

In this expression,

$Y_{i,j}$ represents the standardized response of student i to course j, standardized by his or her responses to the same item for other courses taken at the time as course i;

$G_{i,j}$ is the standardized grade of student i in course j, standardized according to the grades that student i had received or expected to receive in other courses taken concurrently with course i;

R_j is a random effect representing the standardized rating of course j by all students who participated in the survey and took course j;

$P_{i,j}$ is the standardized value of the prior interest of student i in course j's subject matter, as measured by item 3 of the DUET survey;

$A_{i,j}$ is the mean grade awarded in course j, standardized by the mean grades in all courses taken by student i at the time of the survey; and

$e_{i,j}$ is random error that accounts for sources of variation that the other variables cannot explain.

---

[1]Prior student interest was measured by students' responses to item 3 of the DUET survey, even though students completed the survey either 12–13 weeks or nearly two semesters after beginning the course in question. For students who completed the survey twice for the same course, the time lag between repetitions did not seem to have a significant effect on their responses, but there is a legitimate concern over whether student responses to item 3 accurately represent student interest in course subject matter before they began the course or at some time thereafter.

The β's in (4.1) are regression parameters (comparable to the \$125 per square foot in the home price example) that must be estimated from survey data.

Standardized variables[2] are used in this analysis for several reasons. First, regression analyses performed on standardized values of the response and explanatory variables lead to a natural interpretation of the estimated regression coefficients (i.e., the β's) as the change in a student's response ($Y_{i,j}$) per unit change in the corresponding explanatory variable ($G_{i,j}$, R_j, $P_{i,j}$, and $A_{i,j}$). That is, a one standard deviation change in, for example, a student's prior interest in course material is associated with a β_P increase in $Y_{i,j}$, the student's standardized rating of that survey item. The magnitudes of the regression coefficients are thus directly comparable.

In addition to improving the interpretability of the regression model, standardization also eliminates the need to account for tendencies of individual students to be more positive or negative in their responses, and it effectively cancels out many of the differences that might otherwise be attributed to individual student factors, like student motivation and ability. Of course, specific variations in student interest and ability due to such factors as subject matter and personality traits of the instructor cannot be entirely accounted for in this way, but standardization at least eliminates general ability and motivation from the equation. Finally, standardizing student grades eliminates the confounding effect of student GPA on students' perceptions of their grades. As mentioned earlier, a grade of B is likely to affect a C student's perception of his or her performance in a course quite differently than it does an A student's.

Details concerning the standardization procedures used and generalized least squares estimation of the regression coefficients appearing in (4.1) are provided in the appendix at the end of this chapter.

---

[2]A standardized variable is a quantity that has first had its mean subtracted from it, and then has been divided by its standard deviation. For example, if three home prices were \$90,000, \$100,000, and \$110,000, their standardized values could be obtained by first subtracting off their mean (\$100,000). This would result in "centered" values of $-\$10,000$, \$0, and \$10,000. The (sample) standard deviation of these three values is \$10,000, so dividing by the standard deviation would result in values of -1, 0, and 1.

With these definitions in hand, the regression coefficients for each of the explanatory variables described above were estimated for each survey item. These estimates are displayed in Table 1 and were based on the survey responses obtained from all students who knew their final course grades at the time they completed the survey (i.e., responses from non-first-year students in the fall survey period, and responses from all students collected during the spring survey period).

To facilitate the interpretation of the regression coefficients, survey items in Table 1 were grouped according to an underlying factor structure estimated from the data. Although not all groups in the table have a clear interpretation, student responses to items within each group tend to have high correlations with one another, and therefore also tend to have similar regression parameters.[3]

For example, consider the second row in Table 1, which corresponds to the item that probed "Instructor concern." The first entry, 0.224, which appears under the column headed "Student Grade," indicates that a one standard deviation increase in the grade received by a student in a course was associated with a 0.224 increase in the standardized response that the student gave on the item regarding instructor concern. To make this illustration more concrete, suppose that a student rated two courses, and received an A in one course and a C in the other. The standard deviation of these grades is 1.4, and they are 2.0 units apart. All other factors being equal, the average difference between the rating this student would give to the course in which he received an A is thus expected to be $0.224 * 2/1.4 = 0.32$ standardized units higher than the response he gave for the course in which he received a C. In terms of raw units of response on the DUET survey, 0.32 standard deviations corresponds to about 0.4 units on a 5-point scale (the average standard deviation of a DUET item was 1.28).

Continuing this example, the regression coefficient for "Course Rating" was 0.514. In real terms, this means that a one standard

---

[3] Factor loadings were computed using the SAS procedure PROC FACTOR. Orthogonal rotations computed using EQUAMAX, PARSIMAX, and VARIMAX produced nearly identical groupings. The first three items in Group 4, which had slightly higher loadings in Group 3, were placed in the fourth group for ease of interpretation.

TABLE 1

Effects of Intervening Variables on DUET Item Responses Using Received Course Grades. Values in this table summarize estimates from the regression model based on received grades. Standard errors appear in parentheses below each estimate. Unless otherwise indicated, all regression coefficients are significant at the .01 level; those significant at only the .05 level are superscripted by *, while insignificant coefficients are superscripted with a 0.

Item	Student Grade G	Class Mean A	Prior Interest P	Course Rating R
Group 1: Instructor interaction	**0.178**	**0.067**	**0.148**	**0.494**
Instructor concern	0.224	0.104	0.089	0.514
	(0.013)	(0.018)	(0.013)	(0.017)
Encouraged questions	0.171	0.137	0.138	0.497
	(0.013)	(0.017)	(0.012)	(0.017)
Instructor enthusiasm	0.170	0.054	0.156	0.509
	(0.012)	(0.017)	(0.012)	(0.017)
Instructor availability	0.151	0.056	0.072	0.474
	(0.013)	(0.018)	(0.013)	(0.017)
Instructor rating	0.231	0.034^0	0.214	0.526
	(0.013)	(0.018)	(0.013)	(0.018)
Instructor communication	0.210	0.040*	0.201	0.484
	(0.013)	(0.017)	(0.012)	(0.017)
Critical thinking	0.045	0.107	0.142	0.544
	(0.013)	(0.018)	(0.013)	(0.018)
Usefulness of exams	0.266	0.027^0	0.132	0.439
	(0.014)	(0.018)	(0.013)	(0.018)
Related course to research	0.131	0.045*	0.190	0.458
	(0.015)	(0.019)	(0.014)	(0.020)
Group 2: Structure	**0.102**	**−0.039**	**0.151**	**0.398**
Relevance of classes	0.092	−0.056	0.193	0.434
	(0.012)	(0.016)	(0.012)	(0.017)
Relevance of assignments	0.073	−0.065	0.162	0.363
	(0.013)	(0.015)	(0.012)	(0.018)

TABLE 1 *(continued)*

Item	Student Grade G	Class Mean A	Prior Interest P	Course Rating R
Instructor organization	0.154	−0.054	0.126	0.539
	(0.013)	(0.018)	(0.012)	(0.018)
Instructor knowledge	0.115	0.009^0	0.160	0.449
	(0.012)	(0.016)	(0.012)	(0.016)
Knew goals of course	0.076	−0.028	0.115	0.204
	(0.009)	(0.011)	(0.009)	(0.015)
Group 3: Satisfaction with progress	**0.272**	**−0.024**	**0.303**	**0.337**
End interest in subject	0.196	0.005^0	0.572	0.250
	(0.011)	(0.013)	(0.011)	(0.015)
Learned course material	0.410	−0.078	0.295	0.296
	(0.012)	(0.014)	(0.011)	(0.016)
Comparative learning	0.185	−0.073	0.313	0.433
	(0.013)	(0.017)	(0.013)	(0.018)
Accuracy of exams	0.338	−0.047	0.170	0.350
	(0.013)	(0.016)	(0.012)	(0.018)
Another course?	0.251	0.031*	0.203	0.383
	(0.012)	(0.015)	(0.012)	(0.017)
Recommend course?	0.250	0.017^0	0.262	0.311
	(0.012)	(0.014)	(0.011)	(0.016)
Group 4: Difficulty	**−0.168**	**−0.129**	**0.046**	**0.531**
Difficulty of course	−0.227	−0.152	0.039	0.595
	(0.012)	(0.018)	(0.012)	(0.019)
Challenging classes	−0.197	−0.093	0.067	0.556
	(0.013)	(0.018)	(0.013)	(0.019)
Challenging assignments	−0.168	−0.068	0.068	0.549
	(0.013)	(0.019)	(0.013)	(0.020)
Hours/week on assignments	−0.073	−0.108	0.062	0.599
	(0.013)	(0.018)	(0.013)	(0.020)
Stringency of grading	−0.319	−0.236	0.023^0	0.345
	(0.012)	(0.015)	(0.012)	(0.017)
Hours/week in class	−0.026*	−0.116	0.014^0	0.540
	(0.011)	(0.017)	(0.011)	(0.017)

TABLE 1 *(continued)*

Item	Student Grade G	Class Mean A	Prior Interest P	Course Rating R
Group 5: Student work	**0.050**	**0.003**	**0.083**	**0.258**
Completion of written	0.032	0.023*	0.043	0.156
	(0.008)	(0.009)	(0.008)	(0.011)
Completion of reading	0.077	−0.044	0.133	0.349
	(0.013)	(0.016)	(0.012)	(0.018)
Class attendance	0.041	0.031*	0.074	0.268
	(0.010)	(0.012)	(0.010)	(0.012)

deviation change in the average rating of instructor concern received from all students was associated with a 0.514 average increase in the standardized response of a single student to this item. Comparing this to the effect of the students' grades, it follows that the grade a student received in a course was, on average, $.224/.514 = 44\%$ as important in determining a student's rating of instructor concern as the consensus rating of the course obtained from other students.

Most of the DUET items in the first group of Table 1 appear to load heavily on a factor that might be labeled "instructor interaction." Averaged over all items in this group, the relative grade that a student received in a course tends to be a better predictor of a student's response to an item than either the student's prior interest in course material or the mean course grade. Perhaps surprisingly, it carries 36% ($=.178/.494$) of the explanatory power of the course rating variable (represented by the coefficients listed in the final column). Thus, for this group of items, students' grades appear to be approximately one-third as important as the consensus course rating estimated from all student responses on an item for predicting an individual student's response to an item.

The effect of prior student interest was also quite substantial, carrying approximately 30% of the explanatory power of the consensus rating variable. This statistic provides further evidence in support of the assertion that course evaluation forms should be adjusted to account for differences in the composition of students across courses before these forms are used for administrative purposes. Such adjustments are particularly important for instructors who teach courses whose clientele is primarily students whose enrollment is mandated by institutional requirements.

In comparison to prior interest, student grade, and course rating, mean course grade appears to be relatively unimportant as a predictor of student responses for items in the first group.

The second group of items seems to be qualitatively associated with students' perception of the structural coherence of a course. For this group of items, prior student interest is the most significant predictor of student response after the consensus course rating, while standardized student grade is a close third. The former carries approximately 38% of the explanatory power of the consensus course rating, the latter, 26%. Mean course grade again does not appear to be substantively important as a predictor of item responses when student grades are included in the equation.

The underlying factor dominant for items in the third group might be interpreted as representing student satisfaction with progress in a course and the measurement of that progress. For this group of items, both standardized student grade and prior interest weigh heavily in predicting student responses. In fact, prior student interest almost exceeds the importance of the consensus course rating as a predictor, having regression coefficients that are, on average, about 90% as large as the consensus ratings by all students. Of course, the magnitude of this result is due in part to the expectedly high correlation between prior student interest and student interest at the end of the course, but still averages about 70% even without this item. Similarly, the coefficient of standardized student grades is 81% of the coefficient of the consensus course rating.

The fourth group of survey questions probed student perceptions of the difficulty of course material, the extent to which students were challenged, and the time commitment required in a course. Not surprisingly, prior interest is not an important explanatory variable for predicting student responses in this group of items, while both mean course grade and student grade do seem to play an important role. Interestingly, this is the only group of items for which mean course grade consistently improves prediction of student responses, and that within this group, increases in mean course grades and increases in individual student grades are associated with *decreases* in the level of course difficulty, the extent to which students were challenged, and the time students spent on course material. These results call into question the validity of the claim made by many professors that higher grades reflect just compensation to students offered in exchange for higher levels of effort.

The final group of items might be thought to measure an underlying factor representing student commitment. For this group of items, none of the explanatory variables predicts well the responses of students. Even the consensus course rating is relatively ineffective within this group, suggesting that students' attendance patterns and work habits are relatively unaffected by otherwise important course and teacher attributes. This finding echos the results reported by Vasta and Sarmiento [VS79] (the experimental studies in Chapter 3), who found that disparities in grading patterns across sections had little effect on student study habits.

Implications for Grading Theories

Taken together, the parameter estimates in Table 1 provide important insights into the validity of several theories that have been proposed to explain the positive correlation between grades and student evaluations of teaching.

The teacher-effectiveness theory, which posits that good teaching leads to more learning, higher grades, and thus higher teacher–course evaluations, is not supported by the pattern of regression coefficients exhibited in Table 1. According to this theory, the consensus course rating variable should account for essentially all of the variation in students' ratings of course attributes. If grades were to have an effect, it would manifest itself through mean course grades—reflecting improved learning by all students—but not at the individual student level. According to the teacher-effectiveness theory, student grades and prior student interest should not be important factors in predicting student responses on the survey. In fact, however, both student grades and prior student interest typically provide between one-fourth and one-half of the predictive power of the consensus course rating variable.

The negative coefficients of mean course grades in three of the five factor groupings provide further evidence against the teacher-effectiveness theory. These coefficients suggest that as mean course grades increase, the quality of the instructional effectiveness decreases, in direct contradiction to the teacher-effectiveness theory. Further evidence of an inverse relationship between mean course grades and student achievement is presented in Chapter 5.

Of course, the failure of the teacher-effectiveness theory to explain the positive correlation between grades and student evaluations of teaching does not mean that students are unable to judge good teaching. For every item displayed in Table 1, the best predictor of an individual student's response was the consensus rating of the item by other students. Clearly, there is a great deal of agreement among students regarding the quality of instruction received. And herein lies the crux of the debate over the use of student evaluations of teaching for administrative purposes. Many educational researchers, administrators, and faculty claim that such agreement implies that student evaluations of teaching are unaffected by noninstructional factors like student grades and prior interest. Despite these claims, the results cited above make it clear that such factors do affect student evaluations of teaching. The problem is disentangling the effects of noninstructional variables from the effects of actual teaching behaviors.

Theories that explain the positive correlation between grades and student evaluations of teaching through the action of intervening variables, like prior student interest or general student motivation, gain only marginal support from the coefficient estimates cited in Table 1. On the one hand, prior student interest played an important role in determining student responses to most items on the DUET survey. Furthermore, the correlation between prior student interest and student grades was 0.19, suggesting that some of the observed correlation between student grades and student evaluations of teaching found in other studies might be explained through the intervention of this variable. On the other hand, the coefficients in Table 1 suggest an additional effect of student grades beyond that accounted for by prior student interest. Indeed, for many items the association with standardized student grades was stronger than it was with the standardized value of prior student interest.

It is also worth reemphasizing that intervening variables that describe student attributes, like student motivation and student aptitude, are unlikely to underlie the association between student grades and item responses. This is so because the effects of such intervening variables are largely eliminated when both the response and dependent variables are standardized within students.

In contrast to the teacher-effective theory, the grade-leniency theory does provide a viable explanation for many of the patterns exhibited in Table 1. According to the grade-leniency hypothesis, students reward teachers who award high grades. Interpreted at the level of an individual student, this mechanism predicts the signs of essentially all of the regression coefficients reported for the student grade variable. However, this hypothesis is not consistent with many of the coefficients estimated for the class mean grade variables. In particular, the negative coefficients reported for class mean grades in Groups 2 and 3 run contrary to the predictions that this theory naturally yields. Accepting this theory as an explanation for the pattern of coefficients observed in Table 1 thus requires a certain degree of cynicism; not only do students reward teachers who give them high marks, but they are either indifferent to the grades given their classmates or, worse still, are appreciative when their peers receive lower grades than they do.

Of all the commonly proposed explanations for the generally observed positive correlation between student grades and student evaluations of teaching, the grade-attribution theory seems to provide the best explanation for the results summarized in Table 1. According to the grade-attribution hypothesis, students attribute success in academic work to themselves, but attribute failure to external sources. Students who receive comparatively low marks in a course are thus likely to criticize teaching, and indeed this is consistent with the observed pattern of coefficients for the student grade variables in Table 1. In addition, the small, but negative, coefficients associated with class mean grades in the second and third groups of survey questions are not unexpected under this hypothesis if it is assumed that students judge their performance in a class relative to their classmates. If it is true that students make such comparisons, then those students earning low marks in stringently graded classes are less likely to view their performance as a failure than are students who earn low marks in classes that are graded leniently. Thus, low class mean grades might ameliorate the impact of low grades, making it less likely that students would attribute personal failure to their instructor.

The preceding analyses were based on survey data collected from students who had already received their final course grades. However, at most institutions students complete course-evaluation forms before receiving final course grades. For these students, expected rather than received grades are the relevant quantity for predicting students' responses on evaluations of teaching.

In order to investigate the effect of expected grades on student responses to the DUET survey, a parallel analysis using expected grades, as measured by item 22 on the survey and survey data collected from first-year students in the fall semester of 1998 was performed. For these students, responses to the survey were obtained before final course grades had been assigned.

Aside from a minor difference in the scoring of D's,[4] the analysis described above for received grades was repeated a second time by substituting expected grades in their place. Estimates based on the revised definition of the student grade variables appear in Table 2.

A comparison of the regression coefficients reported in Tables 1 and 2 shows that the effects of each of the four explanatory variables on the item responses are nearly identical in the two analyses. The consistency of these results over expected and received grades is reassuring and suggests that the effect of grading on student evaluations of teaching is largely unaffected by the timing of the collection of course evaluation forms.

The extent to which nonresponse may have affected the regression analyses above is discussed in the appendix to this chapter.

---

[4]Due to limitations in the display of response categories on the DUET website, all grades of D (D−, D+, D) were collapsed into a single category.

TABLE 2

Effects of Intervening Variables on DUET Item Responses Using Expected Course Grades. Values in this table summarize estimates from the regression model based on expected grades. Like Table 1, standard errors appear in parentheses below each estimate. Unless otherwise indicated, all regression coefficients are significant at the .01 level; those significant at only the .05 level are superscripted by *, while insignificant coefficients are superscripted with a 0.

Item	Student Grade G	Class Mean A	Prior Interest P	Course Rating R
Group 1: Instructor interaction	**0.182**	**0.067**	**0.148**	**0.520**
Instructor concern	0.216	0.130	0.065	0.522
	(0.026)	(0.030)	(0.024)	(0.031)
Encouraged questions	0.196	0.104	0.162	0.536
	(0.027)	(0.030)	(0.025)	(0.032)
Instructor enthusiasm	0.157	0.039^0	0.184	0.543
	(0.025)	(0.030)	(0.024)	(0.033)
Instructor availability	0.171	0.079	0.068	0.474
	(0.027)	(0.031)	(0.025)	(0.034)
Instructor rating	0.204	0.004^0	0.221	0.569
	(0.026)	(0.030)	(0.025)	(0.031)
Instructor communication	0.219	0.016^0	0.196	0.555
	(0.025)	(0.029)	(0.024)	(0.031)
Critical thinking	0.049^0	0.097	0.157	0.532
	(0.027)	(0.030)	(0.025)	(0.033)
Usefulness of exams	0.245	0.065*	0.135	0.431
	(0.027)	(0.030)	(0.025)	(0.037)
Related course to research	0.161	0.043^0	0.164	0.507
	(0.031)	(0.034)	(0.028)	(0.041)
Group 2: Structure	**0.072**	**−0.053**	**0.174**	**0.475**
Relevance of classes	0.052*	−0.089	0.226	0.477
	(0.025)	(0.028)	(0.023)	(0.034)
Relevance of assignments	0.027^0	−0.058*	0.180	0.387
	(0.026)	(0.029)	(0.024)	(0.035)

TABLE 2 *(continued)*

Item	Student Grade G	Class Mean A	Prior Interest P	Course Rating R
Instructor organization	0.125	-0.048^0	0.139	0.517
	(0.025)	(0.030)	(0.024)	(0.031)
Instructor knowledge	0.084	-0.018^0	0.149	0.517
	(0.024)	(0.029)	(0.023)	(0.031)
Knew goals of course	0.071	-0.048^*	0.114	-0.011^0
	(0.019)	(0.019)	(0.017)	(0.161)
Group 3: Satisfaction with progress	**0.259**	**−0.049**	**0.308**	**0.250**
End interest in subject	0.201	-0.007^0	0.554	−0.292
	(0.022)	(0.023)	(0.021)	(0.032)
Learned course material	0.407	−0.098	0.276	0.268
	(0.024)	(0.025)	(0.022)	(0.040)
Comparative learning	0.155	0.113	0.314	0.439
	(0.028)	(0.029)	(0.026)	(0.039)
Accuracy of exams	0.301	-0.049^0	0.174	0.382
	(0.026)	(0.030)	(0.025)	(0.038)
Another course?	0.231	0.021^0	0.221	0.455
	(0.025)	(0.027)	(0.024)	(0.031)
Recommend course?	0.269	-0.027^0	0.266	0.349
	(0.024)	(0.026)	(0.021)	(0.036)
Group 4: Difficulty	**−0.247**	**−0.103**	**0.075**	**0.512**
Difficulty of course	−0.316	−0.092	0.073	0.540
	(0.026)	(0.031)	(0.025)	(0.037)
Challenging classes	−0.265	−0.083	0.111	0.531
	(0.027)	(0.030)	(0.024)	(0.035)
Challenging assignments	−0.264	-0.021^0	0.074	0.508
	(0.029)	(0.033)	(0.026)	(0.042)
Hours/week on assignments	−0.104	−0.107	0.063^*	0.556
	(0.028)	(0.032)	(0.026)	(0.038)
Stringency of grading	−0.286	−0.212	0.054^*	0.423
	(0.025)	(0.028)	(0.023)	(0.036)
Hours/week in class	-0.039^0	−0.142	0.050^*	0.544
	(0.023)	(0.027)	(0.021)	(0.028)

TABLE 2 *(continued)*

Item	Student Grade G	Class Mean A	Prior Interest P	Course Rating R
Group 5: Student work	**0.076**	**0.003**	**0.100**	**0.266**
Completion of written	0.083	0.038*	0.047	0.170
	(0.015)	(0.016)	(0.013)	(0.021)
Completion of reading	0.070	-0.032^0	0.152	0.362
	(0.027)	(0.030)	(0.024)	(0.040)
Class attendance	0.076	0.033^0	0.084	0.261
	(0.020)	(0.022)	(0.019)	(0.024)

Based on available demographic variables, survey nonresponse does not appear to have had a significant impact on the findings cited above.

The conclusions from these analyses can be summarized as follows. Student responses to the survey were significantly affected by the grades that the students either expected to receive or already had received. For most items, the influence that students' grades had on their responses to the survey ranged from about one-fourth to one-half of the importance of the consensus rating variable estimated from responses collected from all students who took the course. This suggests that although the consensus opinion of instructional attributes was the most important predictor of students' responses to an item, grades do, in fact, represent a serious bias to student evaluations of teaching.

Among the commonly proposed explanations for this bias, the most plausible is provided by the grade attribution theory. It seems that students measure their success in a course, and implicitly the quality of instruction, by the grade they receive or expect to receive. Poor grades are thus associated with poor teaching; students who receive low grades tend to denigrate instruction when they complete teacher–course evaluation forms.

CAUSAL EFFECTS OF STUDENT GRADES ON SETS

I n this section, the impact of expected grades, received grades, and differences between expected and received grades on student evaluations of teaching are examined directly using DUET survey data collected from first-year students at Duke University in the fall semester of 1998 and spring semester of 1999. Because this subset of the DUET data was collected by administering the survey twice to the same group of students, once before and once after they had received their final course grades, it can be used to unambiguously assess the magnitude of the effects of student grades on student evaluations of teaching.

Before describing the analyses of these data, a comment concerning the way in which the survey data were collected is warranted. In order to minimize the possibility that students would discover the purpose for having them complete the same survey for the same courses twice, participating first-year students were told in the fall that they had been asked to complete the survey for their current courses because they, unlike upperclassmen, had not taken courses the previous semester. In the spring, they were again asked to complete the survey for their fall courses, this time under the pretext that this was the default procedure used for all students. Since only one student utilized the survey's comment facility to complain about the "inefficiency" of this procedure, it appears that this deception was effective.

Pairs of student responses collected before and after students had received their final course grades were analyzed using ordinal regression models as described in general terms in Johnson and Albert [Joh97], and in this specific context in Johnson [Joh02]. Although the details of this methodology are too involved to describe here, the basic idea is to estimate the effects of expected and received grades on student responses to the survey by examining changes in item responses that accompanied changes in either their expected or received grades.

Heuristically, the process goes like this. Suppose you want to estimate the relative influence that receiving an A versus a B has on an average student's response to a particular survey item. To accomplish this, you might first group students according to the grade that they said they expected to receive during the fall survey period. Then for each of these groups (e.g., all students expecting to receive an A−), you could compare the average difference in the fall and spring responses collected from those students who ultimately received an A to the average difference between the fall and spring responses collected from those students who ultimately received a B. If, for example, responses collected from the students who ultimately received A's were, on average, 0.5 units higher in the spring than in the fall (when they were expecting to receive an A−), and responses collected from the students who ultimately received B's were an average of 0.5 units lower in the spring than in the fall, you could conclude that the relative effect of receiving an A versus a B, for those students initially expecting an A−, was to increase ratings by 1.0 unit.

The direction of this analysis can also be reversed; changes in student responses for students who received the same grade in the spring can be examined according to the grades they expected to receive in the fall. And in fact, similar comparisons can be made for each combination of expected and received grades in a more formal context using ordinal regression models, as referenced above.

As it happens, results obtained from fitting ordinal regression models to these data are most naturally summarized in terms of odds and odds ratios. For nongamblers, the odds that something occurs is simply the probability that it does occur divided by the probability that it doesn't. An odds ratio is, not surprisingly, the ratio of two odds.

The odds of interest here involve ratios of the probabilities that a particular student rated a survey item higher than, say, category m to the probability that the student rated it in category m or lower. For example, consider the item that probed instructor concern. This item was phrased, "The instructor's concern for the progress of individual students was (1) Very Poor, (2) Poor, (3) Fair, (4) Good, (5) Very Good, (6) Excellent, or (7) Not Applicable." The odds that a student rated a course as being very good or excellent ($m = 4$), can be expressed mathematically as

$$\frac{\textbf{Prob}[\text{response} > 4]}{\textbf{Prob}[\text{response} \leq 4]}.$$

For purposes of the analyses below, pairs of responses containing a "Not applicable" response were ignored.

   Continuing this example, consider the ratio of the odds that on the fall survey, a student who expected to receive an A rated the item higher than category m to the odds that a student who expected to receive a B did. Mathematically, this odds ratio can be expressed

$$\frac{\textbf{Prob} (\text{response} > m, \text{expecting A})}{\textbf{Prob} (\text{response} \leq m, \text{expecting A})} \bigg/ \frac{\textbf{Prob} (\text{response} > m, \text{expecting B})}{\textbf{Prob} (\text{response} \leq m, \text{expecting B})}.$$

$$(4.2)$$

According to the fit provided by the ordinal regression model, this odds ratio can be calculated as

$$\exp(E_A - E_B), \qquad (4.3)$$

where E_A and E_B are the estimated effects that the expectations of an A and B have on student responses to this survey item. These effects are listed in Table 4 in the appendix at the end of this chapter. Survey items appear as rows in this table, and expected grades appear as columns. From Table 4, we see that $E_A = 3.57$ and $E_B = 3.02$. Thus, the numerical value of the odds ratio in (4.2) is

$$\exp(3.57 - 3.02) = 1.73,$$

meaning that the odds that a student expecting an A rates instructor concern higher than category 4 was 1.73 times greater than the odds that a student expecting a B would. In fact, this odds ratio, 1.73, holds for any category of response chosen for this item. That is, the odds that a student expecting an A rated instructor concern higher than "fair" were also 1.73 times greater than the odds that a student who expected a B would. Similarly, the odds that a student expecting

TABLE 3

The probabilities listed in the right column of this table represent the proportion of students who, when expecting to receive an A in a course, would rate the item listed in the left column "highly," under the assumption that 30% of students expecting a B would. A "high" rating simply means higher than any specified category of response. The order of category responses is provided in the DUET survey description appearing in Chapter 2.

Item	Probability
Group 1: Instructor interaction	
Instructor concern	0.43
Encouraged questions	0.45
Instructor enthusiasm	0.40
Instructor availability	0.38
Instructor rating	0.45
Instructor communication	0.50
Critical thinking	0.28
Usefulness of exams	0.41
Related course to research	0.44

an A rated instructor concern higher than "good" were 1.73 times greater than the odds that a student who expected B would, etc.[5]

This illustration can be extended directly to probabilities of response. For example, suppose that there is a 30% probability that a student who expects a B in a course rates the item that probed instructor concern as "good" or higher (i.e., $m \geq 4$). Then the same student, if he or she had expected an A rather than a B, would, on average, rate instructor concern "good" or higher with probability 43%. Clearly, this is a very substantial effect. Corresponding probabilities for the other DUET items are provided in Table 3.

---

[5]The particular type of ordinal regression model described above is called a proportional odds model, and has the property that the values of odds ratios estimated from the model are independent of the particular category m considered. The estimated odds ratio between students expecting an A and students expecting a B on this survey item is thus 1.73 regardless of whether we examine the odds that students rated instructor concern higher than "poor," higher than "fair," higher than "good," or higher than "very good." Theoretical justification for this model can be found in [McC80].

TABLE 3 *(continued)*

Item	Probability
Group 2: Structure	
Relevance of classes	0.47
Relevance of assignments	0.46
Instructor organization	0.40
Instructor knowledge	0.40
Knew goals of course	0.58
Group 3: Satisfaction with progress	
End interest in subject	0.59
Learned course material	0.70
Comparative learning	0.35
Accuracy of exams	0.58
Another course?	0.52
Recommend course?	0.62
Group 4: Difficulty	
Difficulty of course	0.06
Challenging classes	0.13
Challenging assignments	0.15
Hours/week on assignments	0.12
Stringency of grading	0.06
Hours/week in class	0.16
Group 5: Student work	
Completion of written	0.33
Completion of reading	0.29
Class attendance	0.41

For data collected in the spring, received rather than expected grades were relevant. Corresponding effects of received grades on the odds that students rated each survey item higher than a given category are listed in Table 5 in the appendix to this chapter. Again considering the item involving instructor concern, the ratio of the odds that a student who had already received an A rated instructor concern highly to the odds that a student who got a B did may be written

$$\exp(R_A - R_B) = \exp(3.38 - 2.11) = 3.56, \qquad (4.4)$$

where R_A and R_B are the corresponding effects of a student having already received an A or B (as listed in the first row of Table 5). In other words, the odds that a student gives an instructor high marks for concern is increased by a factor of more than 3 if the student gets an A in the course rather than a B!

With so many estimates of grade effects reported in Tables 4 and 5, it is difficult to discern patterns in the listed numerical values of the regression coefficients. However, several interesting trends can be detected through the use of graphical displays like those presented in Figure 1. In this figure, the coefficients from Table 4, representing the effects of expected grades on survey responses, for items in the first factor group are displayed as a function of increasing expected grade.

One striking aspect of the plots in Figure 1 is the nonlinearity of the trend at the lowest grade levels. Apparently, students who expect to receive a grade of D+ or lower are less likely to attribute their poor performance to the instructor than are students who receive grades in the C range. A plausible explanation for this phenomenon, which manifests itself for most of the items in the other groups as well, is that students who expect to receive grades of D+ or lower have not adequately invested themselves in their courses. Keeping in mind that fewer than 3% of all grades assigned to first-year students at Duke are lower than a C−, and that essentially any level of student commitment would earn a grade of at least a C− in a vast majority of Duke courses, it is likely that students who expect to receive such low grades recognize that their poor performance is due to a lack of effort. In other words, students may be willing to attribute poor performance to themselves if such performance reflects an almost total lack of commitment.

The nonlinearity of the effect of expected grade at the low end of the grade spectrum also casts a shadow over the observational studies reviewed in Chapter 3 (as well as the analyses of the previous section!). Because correlation analyses and other linear statistical

FIGURE 1

Comparative effects of expected grade on student evaluations of teaching for items in group 1. Note that only differences in estimated coefficients are relevant for predicting odds ratios.

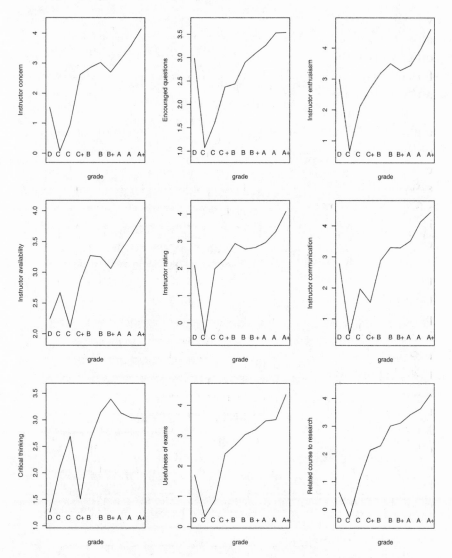

methods employed in these studies were premised on the implicit assumption of a linear relationship between the effects of grades and student evaluations of teaching, the nonlinearity of this relationship raises questions about the validity of results based on these analyses. At the very least, such nonlinearity attenuates the reported relationship between grades and teacher–course evaluations. In more extreme cases, it is plausible that it might completely mask it.

Even more striking than the nonlinearity of the trend in the grade effects at the lowest grade levels is the monotonicity of this trend at more usual grade levels. For eight out of the nine questions in this group of items, the tendency of students to rate more highly those courses for which they received higher grades is nearly uniform. The exception occurs for the item involving the extent to which the instructor required critical thinking, which levels out above grades of B. Because this is the only question in this group that tangentially involves course difficulty, the explanation for this deviation might involve a balancing of the presumably negative effect of the use of stringent grading as a tool to require critical thinking and the positive, attributional effect of receiving higher marks.

The impact of expected grades on student evaluations of teaching for items in the second factor group is less patterned, as Figure 2 demonstrates. In general, however, higher evaluations accompany higher grade expectations, although such effects are far from monotonic. As in the first group, the lowest evaluations were again generated by students who expected to receive C grades.

The effects of expected grades on student evaluations of teaching for items in the third factor group are illustrated in Figure 3. Responses to items that involved interest in subject matter at the end of the course, amount learned, the effectiveness of exams and quizzes, and whether the course would be recommended to others all increased with expected grade. Such trends are well predicted by the grade-attribution and grade-leniency theories. However, there is a telling contrast between the item that queried the amount learned in a course and the item that probed the amount learned in a course compared to other courses taken by the student. The former increases monotonically with grade, while the latter is nearly flat for

FIGURE 2

Comparative effects of expected grade on student evaluations of teaching for items in group 2.

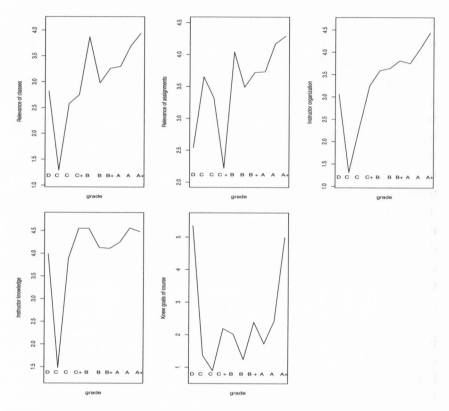

expected grades above B+. The most straightforward explanation of this difference is that grade attribution plays a strong role in determining whether a student felt that he or she had adequately mastered course material, but that courses in which comparatively more learning occurred tended to be graded more severely.

Not surprisingly, over the range of grades normally assigned to Duke students, the final three items in this group exhibit a negative correlation between student ratings and expected grade. These items involve course difficulty and the extent to which lectures and assignments were challenging. When students felt they had performed well

FIGURE 3

Comparative effects of expected grade on student evaluations of teaching for items in group 3.

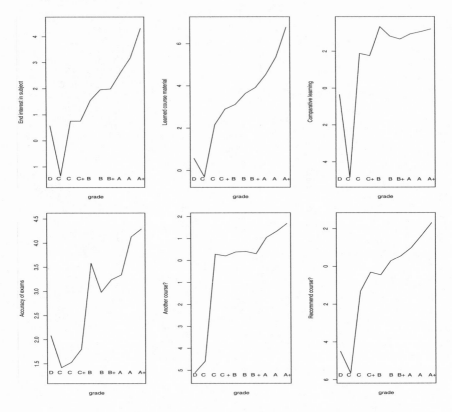

in their courses, as measured by grade expectation, they tended to view their courses as being less difficult and less challenging.

In terms of the effects of expected grades on the ratings of items that probed number of hours per week spent on assignments, stringency of grading, and hours per week spent in class, the questions in the fourth factor group appear similar to the final three questions from the third factor group. A negative trend between ratings and expected grade exists also for these items, which, as before, might be attributed to a perception of diminished course demand among students expecting to receive high marks.

FIGURE 4

Comparative effects of expected grade on student evaluations of teaching for items in group 4.

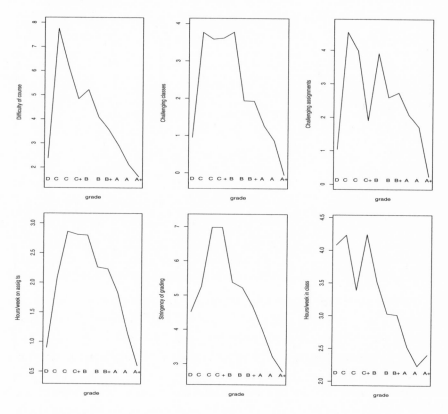

The first two items in the last factor group provide an interesting complement to items in the previous group. Despite the negative association between expected grade and hours per week spent in class and on assignments, there is a positive correlation between expected grade and the proportion of written and reading assignments completed. Students apparently associate high grades with reductions in the time required for a course and with increases in the proportion of reading and written assignments that they were able to complete. Could it be that courses with high grade distributions tend to require less work, thus permitting students to simultaneously

FIGURE 5

Comparative effects of expected grade on student evaluations of teaching for items in group 5.

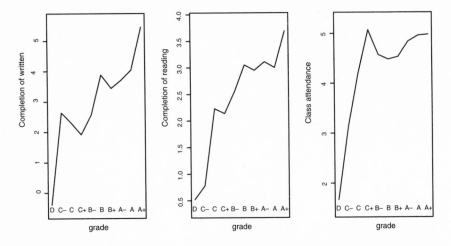

spend less time on such courses while still completing all of their assignments?

Plots comparable to Figures 1 through 5, but based on received grade, are displayed in Figures 6 through 10. Qualitatively, the two sets of plots are commensurate. One feature of these plots is, however, more pronounced in the plots based on received grade: the differences that accompany changes in student evaluations between students receiving a C+ and those receiving a B−. For many items, this difference results in a notable spike in the otherwise dominant trend. This spike is likely caused by the relatively large perceptual difference among undergraduates between the value of a C grade and a B grade.

FIGURE 6

Comparative effects of received grade on student evaluations of teaching for items in group 1.

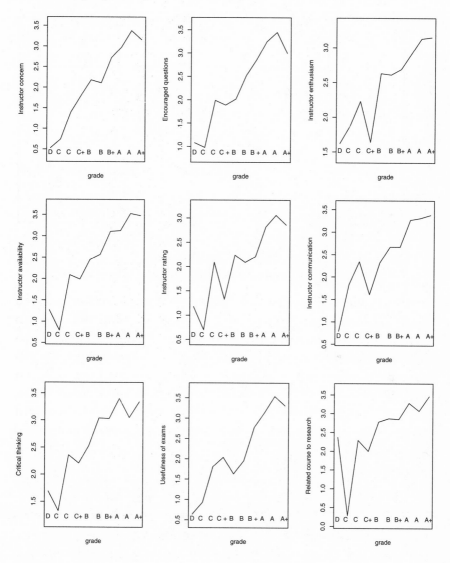

FIGURE 7

Comparative effects of received grade on student evaluations of teaching for items in group 2.

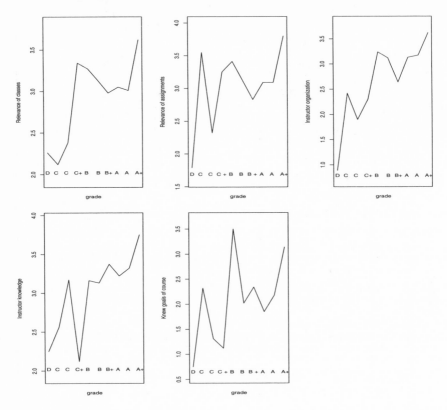

SUMMARY

The DUET experiment was designed so that analyses based on collected data could be controlled for many of the extraneous variables that had disguised the causal relationship between grades and teacher–course evaluations in previous studies. Two analyses were presented to expose this relationship. In the first, standardized responses of students to DUET items were regressed on standardized student grades, standardized values of student-reported prior interest in course subject matter, standardized mean class grades, and standardized values of the consensus rating of an

FIGURE 8

Comparative effects of received grade on student evaluations of teaching for items in group 3.

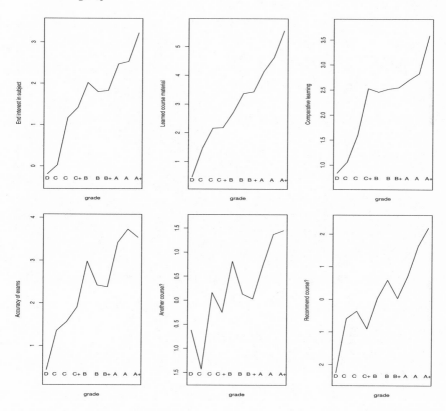

item by other students in the course. The purpose of standardization was to eliminate the effects of many of the nuisance variables previously mentioned; that is, standardization effectively removed biases that might otherwise be attributed to student aptitude, gender, race, academic major, academic year, prior grade expectation, and differing levels of general student interest. Standardization of variables also facilitated the interpretation of the relative importance of each variable in predicting student responses on the survey items.

The results of the regression analyses varied across survey items. Generally speaking, however, standardized student grade was approximately one-quarter to one-half as important for pre-

FIGURE 9

Comparative effects of received grade on student evaluations of teaching for items in group 4.

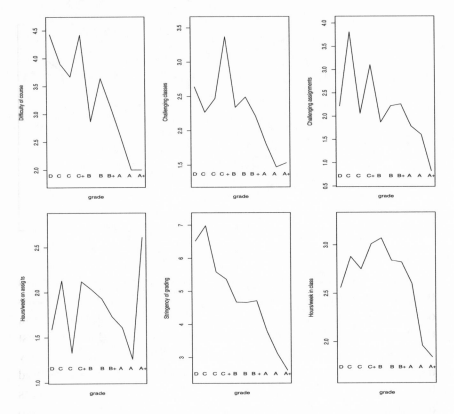

dicting student responses to the survey as was the standardized value of the consensus rating obtained from all students. Prior student interest had effects that were similar to student grades. Furthermore, as demonstrated in the appendix below, the effects of prior student interest and student grade were relatively consistent across students from different academic disciplines, ethnic groups, genders, and academic abilities. Finally, class mean grade was generally an order of magnitude less important as a predictor of item responses than were either individual student grades or prior student interest.

FIGURE 10

Comparative effects of received grade on student evaluations of teaching for items in group 5.

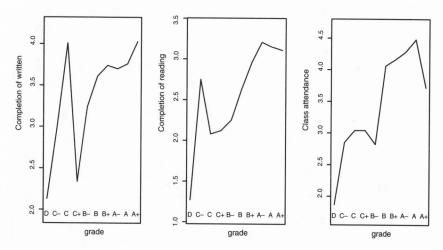

Of the four major theories that have been proposed to explain the positive correlation between grades and student evaluations of teaching, the grade-attribution theory and, to a lesser extent, theories based on the operation of intervening variables are most useful. Little support was garnered for the grade-leniency theory because class mean grades, unlike individual student grades, had little influence on student responses on the survey. The strong effects of both standardized student grades and prior interest variables also contradict predictions generated under the teacher-effectiveness theory. Under that hypothesis, effective teaching should manifest itself primarily through the consensus course rating variable. This is, of course, not to say that students were unable to discern effective teaching; in fact, the consensus rating variable was the most important variable in predicting student responses to the survey. However, the additional predictive power of both student grade and prior student interest suggests that teacher/course attributes alone are not responsible for the generally observed positive correlation between student evaluations of teaching and student grades.

By using a subset of the DUET data collected from first-year students, a more quantitative investigation of the effects of student grades on teacher–course evaluations was performed. The unique feature of this analysis was that two student responses were obtained for each student–course combination. The first response was obtained while students were still taking their courses and before they had received their final course grades. The second response was collected after students had completed their courses and after they had received their final course grades. By contrasting the two responses obtained from each student, highly significant, substantively important effects of student grade were discovered on all of the DUET items. These findings corroborate the findings of earlier grade-manipulation studies and a preponderance of correlational studies. Because the design of the DUET experiment effectively eliminated the possibility that unobserved environmental factors were responsible for these effects, the results from this analysis provide conclusive evidence of a biasing effect of student grades on student evaluations of teaching.

Appendix

Standardization Procedures for Analyses of "Intervening Variables and SETS"

Details of the methodology used to standardize the variables appearing in (4.1) to obtain least squares estimates of β_G, β_R, β_P, and β_A are as follows.

To begin, consider the response variable $Y_{i,j}$, the standardized response of student i on a given DUET item for course j, and let $y_{i,j}$ denote the corresponding observed (nonstandardized) response of student i for course j. Typically, $y_{i,j}$ is an ordinal variable coded from, say, 1 to 5, corresponding to the ordered categories ranging from, say, "very bad" to "excellent." Then the standardized response variable $Y_{i,j}$ was defined as

$$Y_{i,j} = \frac{y_{i,j} - \bar{y}_i}{s_{y_i}},$$

where $\bar{y}_i$ is the average response by student i to the given survey item, averaged over all the responses that student i provided for courses taken concurrently with course j. The quantity s_{y_i} represents the standard deviation of responses by student i on this same set of courses. Specifically, if student i completed the survey for n_i courses, then

$$\bar{y}_i = \frac{1}{n_i} \sum_{j=1}^{n_i} y_{i,j} \quad \text{and} \quad s_{y_i}^2 = \frac{1}{n_i - 1} \sum_{j=1}^{n_i} (y_{i,j} - \bar{y}_i)^2.$$

Relative grades of students in their courses were defined similarly, using either received grades or expected grades. Received grades were determined from official student records, while expected grades were defined from student responses to item 22 of the survey. If $g_{i,j}$ represents the grade received (or expected) by student i in course j (coded $A = 4.0, A- = 3.7, B+ = 3.3, \ldots, D = 1.0, F = 0.0$), then $G_{i,j}$ was defined as

$$G_{i,j} = \frac{g_{i,j} - \bar{g}_i}{s_{g_i}},$$

where $\bar{g}_i$ and s_{g_i} represent the mean grade and standard deviation of the grades received by student i in courses that he or she rated during the survey period.

The consensus course rating variable R_j represents an estimate of the average rating of course j obtained from all students who took course j. A priori, the distribution of R_j was assumed to have mean zero and standard deviation one. Unlike the other explanatory variables, the values of R_j were not directly observable from the DUET data. Instead, the values of these variables were estimated jointly with the regression coefficients and other model parameters. Such variables are examples of "random effects." Because the other explanatory variables were standardized in the regression model, the prior distributions assumed for the values of R_j were assumed to be independent standard normal distributions.

The standardized mean course grade, denoted by $A_{i,j}$, was computed as

$$A_{i,j} = \frac{a_{i,j} - \bar{a}}{s_a},$$

where $a_{i,j}$ denotes the mean grade of all students who took course j with student i, and

$$\bar{a} = \frac{1}{n_i} \sum a_{i,k} \quad \text{and} \quad s_a^2 = \frac{1}{n_i - 1} \sum (a_{i,k} - \bar{a})^2.$$

Both sums extend over courses k taken by student i.

The final explanatory variable included in the regression equation was the standardized response of each student to the survey item that questioned prior interest. This variable was standardized in the same manner as $Y_{i,j}$, and was denoted by $P_{i,j}$.

The error terms in the regression equations, $(e_{i,j})$, were assumed to be normally distributed, but were not assumed to be independent because of the standardization of responses within students. A more general form of the covariance matrix was used to account for the $1/(n_i - 1)$ correlation between the standardized values computed for each student. Details on estimation for linear models with dependent observations and singular covariances may be found in, for example, [Rao73], while details concerning computational methods for estimating parameters in linear models with random effects may be found in [GHRPS90].

Effects of Nonresponse in Analyses of Intervening Variables and SETs

As discussed in Chapter 2, only about 1 in 5 students participated in the DUET survey in each of the semesters the experiment was conducted. This raises questions as to the extensibility of conclusions to student groups who did not participate in the survey.

One avenue available for investigating the effects of nonresponse is to examine interactions between the explanatory variables appearing in (4.1) and demographic variables associated with students' response patterns. Five such variables were identified in Chapter 2: gender, ethnicity (American Indian, Asian, African-American, Hispanic, White, or other), academic year (freshman, sophomore, junior, or senior), GPA, and academic division (humanities, social science, natural science and mathematics, and engineering majors). To test whether the effects of the explanatory variables differed within these demographic groups, interaction terms for all combinations of each demographic variable with both standardized student grade and class mean grade were estimated using maximum likelihood procedures.[6] In this context, the interaction terms allow regression coefficients to be estimated separately for each demographic group. Interaction terms that are significantly different from zero imply that the effect of an explanatory variable is different for the corresponding demographic group.

The total number of interaction models fit was $29 \times 5 \times 2 + 1 = 291$, one for each of the 29 items, 5 demographic variables, and 2 explanatory variables, plus one for the null model with no interactions. The significance of the interaction terms in each model was assessed using likelihood-ratio tests. All tests were evaluated at the 5% level of significance using Bonferroni adjustments.[7] Perhaps surprisingly, none of the 290 interaction terms were found to be statistically significant using this procedure.

Of course, with so many interaction models to test, the power of each test to detect significant interactions is greatly diminished when Bonferroni adjustments are made. Because of this lack of power, the 29 significance tests conducted for each demographic variable were also reevaluated separately using the Bonferroni correction for only the 29 items tested for each demographic variable.

---

[6]Maximum likelihood estimates of the random effects were constrained to have unit variance. Also, because freshman at Duke University do not declare majors, only the data underlying the analyses summarized in Table 1 were used to examine the interaction terms.

[7]Under the classical statistics paradigm, significance tests are defined so that the probability that the null hypothesis (i.e., that the interaction terms are negligible) is rejected occurs in only, say, 5% of the cases in which it is actually true. However, if two or more tests are conducted and the null hypothesis is rejected at the 5% level in each, then there is a greater than 5% chance that at least one valid null hypothesis will be rejected. Bonferroni adjustments are used to guarantee that no valid null hypothesis will be rejected more than, say, 5% of the time in any of the multiple tests considered. The disadvantage of Bonferroni adjustment is that it makes it more difficult to reject the null hypotheses in all of the tests, even when some of them are not valid.

Even at this reduced level of significance, none of the interaction terms between student GPA and either standardized student grade or class mean grade were significant at the 5% level for any of the 29 DUET items appearing in Table 1. Similarly, the gender–student grade and academic division–class mean grade interactions were not significant for any of the 29 items. In nonstatistical terms, this means that the effect of standardized student grade and class mean grade for predicting item responses was approximately independent of students' GPA, that the effect of standardized student grade in predicting item responses was approximately independent of gender, and that academic division did not significantly alter the predictive effect of class mean grade.

Two significant interactions were, however, detected between academic division and student grade. The items for which this interaction was significant were "How easy was it to meet with the instructor outside of class?" and "How would you rate the organization of the instructor(s) in this course?" For the first question (instructor availability), the estimated regression coefficients for student grade, when estimated by academic division, were 0.102 for social science majors, 0.186 for engineering majors, 0.131 for humanities majors, and 0.211 for science majors. For the second item, the estimated regression coefficients for student grade were 0.105 for social science majors, 0.241 for engineering majors, 0.171 for humanities majors, and 0.200 for natural science majors. Weighting these coefficients according to the proportion of each major at Duke during this period resulted in regression coefficient estimates of 0.145 and 0.161 for the student grade variable for these two items. From Table 1, the corresponding unweighted estimates were 0.151 and 0.154, respectively.

Similarly, two possibly important interactions between gender and class mean grade were detected. For the question "How well did you learn and understand course material," the regression coefficient of class mean grade for male students was -0.077, while the estimated coefficient for females was -0.102. This might imply that the negative effect of class mean grade on female students was stronger than for males on this item, though it should be noted that the regression coefficients for males and females for the related item

"How much did you learn in this course compared to all courses that you have taken at Duke?" were nearly identical. In any case, if this difference is not simply the result of random variation, then the corrected estimate of the regression coefficient for class mean grade on this item, corrected for the known response probabilities of female and male students, is -0.089. For comparison, the uncorrected value is -0.078. Other regression coefficients in the model were not substantially affected by the inclusion of this interaction term.

The second interaction detected between gender and class mean grade occurred for the item "What proportion of the written assignments did you complete?" For this item, the estimated regression coefficients for males and females were 0.045 and 0.013, respectively. Because the predictive power of class mean grade is only marginally significant for this item anyway, the difference between these two estimates was not judged to be substantively important.

Ethnicity exhibited a moderate interaction with student grade on the item "How would you rate the organization of the instructor(s) in this course?" The regression coefficients associated with Native Americans, Asians, and African-Americans were larger than those associated with Hispanics, Whites, and Other categories. The adjusted regression coefficient of student grade, adjusted for the observed response probabilities for each demographic group, was 0.141, as compared to the unadjusted estimate of 0.154.

The interaction between ethnicity and class mean grade was also marginally significant for the item that probed instructor concern for students. For that item, the regression coefficient for Native American students was significantly smaller than those for the other groups (0.019), while the effect for Asian-American students was significantly higher (0.375). The regression coefficients for the other groups were similar, and were close to the nonresponse-corrected estimate of 0.275.

Academic year appeared to be the most significant demographic variable in terms of its influence on the explanatory variables of student grade and class mean grade. Overall, five interactions between academic year and the two grade variables were identified.

Two of these five interactions represented modifications of the effects of class mean grade. For the item that probed instructor availability, the regression coefficient for class mean grade was largest for seniors and decreased monotonically for juniors, sophomores, and freshmen. The adjusted estimate of the effect of class mean grade was 0.026, as compared to the unadjusted estimate of 0.056. The interaction of academic year with the item concerning students' interest at the end of the course was less patterned, with seniors and freshmen exhibiting smaller effects than juniors and sophomores. The adjusted estimate of the effect of class mean grade for this item was −0.013, while the unadjusted estimate from Table 1 was 0.005.

Academic year may have also significantly modified the effect of student grade in predicting items involving instructor organization, learning course material, and student reported hours per week attending class meetings, discussion sessions, and labs. The most unusual result in this regard involves the estimated interaction of standardized student grade with first-year status for the item that probed instructor organization. In this case, the effect of standardized student grades had almost the opposite effect in predicting freshman ratings of instructor organization as they did for sophomores, juniors, and seniors. However, the variance of this interaction term was comparatively large, and the estimate of the interaction term for freshmen was just significantly different from zero at the 5% level (without a Bonferroni correction). Nonetheless, the adjusted estimate of this coefficient, adjusted for academic year, was 0.122, while the unadjusted estimate was 0.154.

The statistically significant interaction between grade and academic year for the item that probed "learning course material" occurred at the sophomore level, for which the effects of student grade appeared larger than for the other years. The adjusted and unadjusted estimates for student grade on this item were similar, 0.379 versus 0.410. Finally, freshmen appeared to be much more influenced by their grades when responding to the item involving the number of hours per week devoted to class attendance. The estimate of the regression coefficient of student grade for this item, adjusted for response probabilities of the various year groups, was 0.007, as compared to an unadjusted estimate of 0.029.

Overall, the effects of nonresponse on the estimates reported in Table 1 appear to be quite weak, at least to the extent that nonresponse was associated with available demographic variables.

Regression Coefficients for Analyses of Causal Effects of Grades on SETs

See Tables 4 and 5 on the following pages.

TABLE 4

Regression Coefficients of Expected Grades. These coefficients may be used in conjunction with equation (4.3) to compute the ratio of the odds that a student expecting one of two grades rates a DUET item higher than any given category m.

Item	E_D	E_{C-}	E_C	E_{C+}	E_{B-}	E_B	E_{B+}	E_{A-}	E_A	E_{A+}
Group 1: Instructor interaction										
Instructor concern	1.53	0.07	0.93	2.62	2.85	3.02	2.70	3.13	3.57	4.14
	(1.03)	(0.55)	(0.28)	(0.42)	(0.20)	(0.16)	(0.14)	(0.13)	(0.14)	(0.21)
Encouraged questions	2.98	1.07	1.61	2.37	2.44	2.90	3.09	3.26	3.53	3.54
	(0.71)	(0.44)	(0.29)	(0.42)	(0.23)	(0.18)	(0.17)	(0.17)	(0.16)	(0.23)
Instructor enthusiasm	2.99	0.67	2.12	2.70	3.19	3.50	3.28	3.44	3.96	4.61
	(0.72)	(0.52)	(0.28)	(0.42)	(0.23)	(0.19)	(0.18)	(0.17)	(0.18)	(0.25)
Instructor availability	2.24	2.67	2.10	2.85	3.27	3.25	3.06	3.35	3.60	3.88
	(0.82)	(0.58)	(0.29)	(0.39)	(0.22)	(0.16)	(0.15)	(0.13)	(0.13)	(0.22)
Instructor rating	2.11	−0.43	1.99	2.34	2.92	2.71	2.77	2.96	3.36	4.10
	(0.79)	(0.72)	(0.31)	(0.40)	(0.22)	(0.16)	(0.14)	(0.13)	(0.12)	(0.22)
Instructor communication	2.78	0.52	1.97	1.54	2.88	3.30	3.29	3.52	4.13	4.44
	(0.70)	(0.51)	(0.29)	(0.39)	(0.24)	(0.21)	(0.21)	(0.20)	(0.21)	(0.26)
Critical thinking	1.26	2.10	2.69	1.51	2.63	3.14	3.39	3.13	3.04	3.03
	(0.73)	(0.66)	(0.29)	(0.41)	(0.22)	(0.17)	(0.16)	(0.14)	(0.13)	(0.22)
Usefulness of exams	1.70	0.33	0.88	2.40	2.69	3.04	3.20	3.49	3.53	4.35
	(0.80)	(0.53)	(0.31)	(0.43)	(0.25)	(0.21)	(0.20)	(0.20)	(0.19)	(0.28)
Related course to research	0.60	−0.29	1.10	2.14	2.30	3.01	3.11	3.42	3.63	4.15
	(0.98)	(0.69)	(0.28)	(0.39)	(0.22)	(0.19)	(0.16)	(0.16)	(0.15)	(0.24)

Group 2: Structure

	1	2	3	4	5	6	7	8	9	10
Relevance of classes	2.82 (0.73)	1.29 (0.47)	2.58 (0.29)	2.75 (0.43)	3.87 (0.22)	2.98 (0.15)	3.26 (0.14)	3.30 (0.12)	3.69 (0.12)	3.94 (0.21)
Relevance of assignments	2.54 (0.73)	3.65 (0.58)	3.32 (0.31)	2.22 (0.52)	4.04 (0.23)	3.49 (0.16)	3.72 (0.15)	3.73 (0.13)	4.17 (0.13)	4.29 (0.23)
Instructor organization	3.06 (0.78)	1.31 (0.57)	2.30 (0.28)	3.25 (0.41)	3.59 (0.24)	3.64 (0.19)	3.81 (0.18)	3.75 (0.17)	4.08 (0.17)	4.44 (0.23)
Instructor knowledge	3.99 (0.89)	1.48 (0.49)	3.89 (0.34)	4.55 (0.46)	4.55 (0.30)	4.13 (0.25)	4.11 (0.24)	4.25 (0.23)	4.56 (0.23)	4.48 (0.27)
Knew goals of course	5.35 (1.96)	1.37 (0.61)	0.90 (0.29)	2.19 (0.53)	2.02 (0.28)	1.24 (0.15)	2.39 (0.19)	1.72 (0.12)	2.42 (0.13)	4.99 (0.48)
Group 3: Satisfaction with progress										
End interest in subject	0.57 (0.65)	−1.35 (0.54)	0.75 (0.24)	0.75 (0.36)	1.55 (0.17)	1.96 (0.11)	1.99 (0.10)	2.61 (0.12)	3.18 (0.11)	4.31 (0.19)
Learned course material	2.57 (0.81)	−0.30 (0.49)	2.15 (0.28)	2.90 (0.39)	3.12 (0.21)	3.64 (0.16)	3.92 (0.15)	4.53 (0.14)	5.36 (0.14)	6.77 (0.22)
Comparative learning	−0.35 (0.82)	−4.83 (1.83)	1.88 (0.31)	1.76 (0.37)	3.34 (0.25)	2.83 (0.16)	2.67 (0.14)	2.95 (0.12)	3.07 (0.12)	3.21 (0.20)
Accuracy of exams	2.08 (0.76)	1.42 (0.49)	1.53 (0.29)	1.80 (0.38)	3.58 (0.22)	2.98 (0.16)	3.24 (0.15)	3.34 (0.13)	4.13 (0.14)	4.29 (0.23)
Another course?	−5.11 (2.23)	−4.59 (1.18)	0.29 (0.32)	0.21 (0.40)	0.39 (0.20)	0.41 (0.13)	0.31 (0.11)	1.05 (0.09)	1.33 (0.10)	1.68 (0.21)
Recommend course?	−4.50 (1.92)	−5.65 (1.45)	−1.32 (0.32)	−0.31 (0.41)	−0.45 (0.21)	0.29 (0.13)	0.56 (0.11)	0.99 (0.09)	1.64 (0.09)	2.31 (0.26)

TABLE 4 (continued)

Item	E_D	E_{C-}	E_C	E_{C+}	E_{B-}	E_B	E_{B+}	E_{A-}	E_A	E_{A+}
Group 4: Difficulty										
Difficulty of course	2.38	7.75	6.18	4.84	5.22	4.09	3.56	2.90	2.12	1.62
	(0.79)	(0.83)	(0.36)	(0.39)	(0.24)	(0.15)	(0.14)	(0.11)	(0.09)	(0.20)
Challenging classes	0.95	3.77	3.59	3.62	3.78	1.94	1.93	1.26	0.87	−0.05
	(0.92)	(0.57)	(0.30)	(0.43)	(0.21)	(0.13)	(0.11)	(0.08)	(0.07)	(0.20)
Challenging assignments	1.04	4.54	3.98	1.90	3.90	2.59	2.73	2.07	1.70	0.23
	(0.92)	(0.54)	(0.33)	(0.60)	(0.24)	(0.15)	(0.13)	(0.10)	(0.09)	(0.20)
Hours/week on assigments	0.90	2.09	2.86	2.81	2.80	2.26	2.23	1.83	1.15	0.60
	(0.98)	(0.47)	(0.29)	(0.41)	(0.20)	(0.14)	(0.11)	(0.09)	(0.08)	(0.19)
Stringency of grading	4.52	5.26	6.98	6.98	5.38	5.22	4.69	3.98	3.22	2.78
	(0.79)	(0.55)	(0.31)	(0.38)	(0.24)	(0.19)	(0.18)	(0.16)	(0.14)	(0.21)
Hours/week in class	4.08	4.23	3.39	4.24	3.51	3.03	3.01	2.52	2.23	2.40
	(0.74)	(0.51)	(0.26)	(0.39)	(0.20)	(0.14)	(0.12)	(0.09)	(0.08)	(0.18)
Group 5: Student work										
Completion of written	−0.39	2.64	2.30	1.92	2.56	3.88	3.44	3.71	4.04	5.45
	(0.76)	(0.65)	(0.35)	(0.48)	(0.28)	(0.28)	(0.21)	(0.19)	(0.18)	(0.60)
Completion of reading	0.52	0.78	2.23	2.14	2.55	3.05	2.95	3.11	3.00	3.69
	(0.91)	(0.51)	(0.31)	(0.44)	(0.24)	(0.16)	(0.13)	(0.12)	(0.11)	(0.24)
Class attendance	1.67	3.13	4.20	5.07	4.57	4.48	4.53	4.84	4.96	4.98
	(0.74)	(0.49)	(0.35)	(0.56)	(0.27)	(0.20)	(0.17)	(0.16)	(0.15)	(0.26)

TABLE 5

Regression Coefficients of Received Grades. These coefficients may be used in conjunction with equation (4.4) to compute the ratio of the odds that a student expecting one of two grades rates a DUET item higher than any given category m.

Item	R_D	R_{C-}	R_C	R_{C+}	R_{B-}	R_B	R_{B+}	R_{A-}	R_A	R_{A+}
Group 1: Instructor interaction										
Instructor concern	0.53 (0.31)	0.74 (0.31)	1.40 (0.21)	1.80 (0.23)	2.18 (0.24)	2.11 (0.13)	2.72 (0.14)	2.98 (0.14)	3.38 (0.13)	3.16 (0.20)
Encouraged questions	1.08 (0.32)	0.98 (0.31)	1.99 (0.21)	1.89 (0.24)	2.02 (0.22)	2.52 (0.12)	2.86 (0.11)	3.25 (0.10)	3.45 (0.10)	3.00 (0.17)
Instructor enthusiasm	1.62 (0.33)	1.87 (0.35)	2.23 (0.24)	1.64 (0.24)	2.63 (0.23)	2.61 (0.16)	2.69 (0.16)	2.91 (0.17)	3.13 (0.16)	3.15 (0.23)
Instructor availability	1.27 (0.30)	0.79 (0.37)	2.09 (0.21)	1.99 (0.24)	2.45 (0.22)	2.57 (0.13)	3.11 (0.11)	3.13 (0.10)	3.53 (0.11)	3.48 (0.19)
Instructor rating	1.18 (0.34)	0.70 (0.34)	2.09 (0.21)	1.33 (0.24)	2.24 (0.22)	2.09 (0.12)	2.21 (0.12)	2.82 (0.11)	3.06 (0.10)	2.86 (0.18)
Instructor communication	0.79 (0.28)	1.83 (0.31)	2.34 (0.22)	1.61 (0.23)	2.32 (0.20)	2.67 (0.12)	2.67 (0.11)	3.27 (0.11)	3.31 (0.10)	3.38 (0.18)
Critical thinking	1.69 (0.35)	1.33 (0.33)	2.36 (0.23)	2.21 (0.24)	2.53 (0.22)	3.04 (0.14)	3.03 (0.14)	3.40 (0.13)	3.05 (0.12)	3.34 (0.21)
Usefulness of exams	0.64 (0.28)	0.93 (0.33)	1.81 (0.22)	2.04 (0.25)	1.63 (0.22)	1.96 (0.13)	2.78 (0.14)	3.14 (0.14)	3.54 (0.14)	3.31 (0.21)
Related course to research	2.37 (0.38)	0.29 (0.53)	2.29 (0.26)	2.00 (0.27)	2.78 (0.24)	2.87 (0.14)	2.85 (0.14)	3.28 (0.11)	3.07 (0.10)	3.46 (0.22)

TABLE 5 (continued)

Item	R_D	R_{C-}	R_C	R_{C+}	R_{B-}	R_B	R_{B+}	R_{A-}	R_A	R_{A+}
Group 2: Structure										
Relevance of classes	2.26 (0.29)	2.12 (0.34)	2.38 (0.22)	3.34 (0.25)	3.27 (0.22)	3.13 (0.14)	2.98 (0.13)	3.05 (0.12)	3.01 (0.11)	3.62 (0.20)
Relevance of assignments	1.79 (0.31)	3.55 (0.36)	2.32 (0.22)	3.25 (0.26)	3.41 (0.21)	3.12 (0.13)	2.83 (0.13)	3.09 (0.12)	3.09 (0.11)	3.80 (0.21)
Instructor organization	0.89 (0.29)	2.42 (0.33)	1.90 (0.21)	2.30 (0.24)	3.24 (0.21)	3.12 (0.13)	2.64 (0.12)	3.13 (0.11)	3.17 (0.10)	3.62 (0.19)
Instructor knowledge	2.25 (0.34)	2.56 (0.32)	3.17 (0.22)	2.12 (0.22)	3.16 (0.20)	3.13 (0.11)	3.37 (0.11)	3.22 (0.09)	3.32 (0.08)	3.75 (0.18)
Knew goals of course	0.75 (0.31)	2.32 (0.57)	1.31 (0.25)	1.12 (0.28)	3.50 (0.52)	2.02 (0.18)	2.34 (0.16)	1.85 (0.12)	2.18 (0.11)	3.14 (0.34)
Group 3: Satisfaction with progress										
End interest in subject	-0.19 (0.27)	0.03 (0.34)	1.17 (0.18)	1.42 (0.23)	2.02 (0.17)	1.80 (0.11)	1.83 (0.09)	2.47 (0.08)	2.53 (0.08)	3.22 (0.15)
Learned course material	0.47 (0.27)	1.47 (0.33)	2.16 (0.21)	2.18 (0.23)	2.71 (0.21)	3.37 (0.16)	3.43 (0.15)	4.14 (0.15)	4.62 (0.15)	5.55 (0.23)
Comparative learning	0.84 (0.33)	1.06 (0.35)	1.60 (0.21)	2.53 (0.26)	2.46 (0.21)	2.52 (0.12)	2.55 (0.12)	2.70 (0.11)	2.83 (0.11)	3.59 (0.19)
Accuracy of exams	0.44 (0.28)	1.36 (0.34)	1.57 (0.20)	1.91 (0.23)	2.98 (0.22)	2.42 (0.13)	2.38 (0.12)	3.42 (0.12)	3.73 (0.12)	3.54 (0.19)

Another course?	−0.62	−1.43	0.16	−0.25	0.81	0.13	0.03	0.73	1.37	1.45
	(0.33)	(0.44)	(0.22)	(0.26)	(0.24)	(0.12)	(0.10)	(0.09)	(0.10)	(0.25)
Recommend course?	−2.25	−0.59	−0.36	−0.90	0.03	0.59	0.03	0.72	1.63	2.19
	(0.53)	(0.37)	(0.26)	(0.29)	(0.20)	(0.11)	(0.11)	(0.09)	(0.09)	(0.28)
Group 4: Difficulty										
Difficulty of course	4.43	3.90	3.67	4.42	2.87	3.64	3.12	2.57	2.00	2.00
	(0.35)	(0.35)	(0.22)	(0.25)	(0.21)	(0.13)	(0.13)	(0.11)	(0.09)	(0.17)
Challenging classes	2.64	2.27	2.47	3.37	2.34	2.49	2.21	1.81	1.47	1.53
	(0.31)	(0.35)	(0.21)	(0.25)	(0.22)	(0.12)	(0.11)	(0.09)	(0.08)	(0.18)
Challenging assigments	2.22	3.81	2.06	3.10	1.87	2.22	2.26	1.78	1.60	0.82
	(0.35)	(0.38)	(0.20)	(0.25)	(0.20)	(0.12)	(0.12)	(0.10)	(0.08)	(0.21)
Hours/week on assigments	1.59	2.13	1.33	2.12	2.03	1.93	1.73	1.61	1.26	1.65
	(0.29)	(0.34)	(0.22)	(0.24)	(0.20)	(0.12)	(0.11)	(0.09)	(0.08)	(0.19)
Stringency of grading	6.52	6.98	5.58	5.36	4.67	4.66	4.71	3.78	3.11	2.61
	(0.30)	(0.32)	(0.22)	(0.24)	(0.22)	(0.14)	(0.14)	(0.12)	(0.11)	(0.18)
Hours/week in class	2.56	2.88	2.75	3.01	3.07	2.84	2.82	2.60	1.96	1.84
	(0.30)	(0.30)	(0.22)	(0.24)	(0.20)	(0.12)	(0.12)	(0.10)	(0.08)	(0.19)
Group 5: Student work										
Completion of written	2.13	3.01	4.01	2.34	3.24	3.61	3.74	3.70	3.76	4.03
	(0.32)	(0.38)	(0.25)	(0.25)	(0.23)	(0.15)	(0.14)	(0.14)	(0.12)	(0.24)
Completion of reading	1.27	2.75	2.08	2.12	2.25	2.63	2.96	3.21	3.15	3.11
	(0.32)	(0.42)	(0.24)	(0.25)	(0.21)	(0.13)	(0.13)	(0.12)	(0.10)	(0.20)
Class attendance	1.87	2.85	3.04	3.04	2.82	4.06	4.16	4.28	4.48	3.72
	(0.29)	(0.33)	(0.24)	(0.25)	(0.24)	(0.17)	(0.17)	(0.16)	(0.15)	(0.22)

Validity of Student Evaluations of Teaching

Student evaluations of teaching dominate the assessment of instructional quality in America's colleges and universities. Yet the connections between student assessments of instructional quality and actual teaching effectiveness are not well established. In this chapter, the strength of these connections is examined and found wanting. Student evaluations of teaching are strongly influenced by a number of factors not related to subsequent student achievement, and are particularly susceptible to biases attributable to instructor enthusiasm, expressiveness, and grading policy. Furthermore, student evaluations of teaching are found to be relatively unaffected by course content.

A proposal to more firmly establish the relation between selected items on student evaluations of teaching forms and student achievement, using the performance of students in follow-on courses as an outcome measure, is also examined.

I N THE TWO PREVIOUS CHAPTERS, INFLUENCES THAT grades exert on student evaluations of teaching were examined. In this chapter, the broader issue of the validity of student ratings of instruction for assessing faculty teaching is studied. Unlike many studies reviewed in the previous chapter, the view of effective teaching adopted here is that measures of effective teaching must be linked to and dependent upon student learning. Accepting such a view immediately calls into question a number of issues pertaining to the use of student evaluations of teaching, including the purposes for which teacher–course evaluations are collected, the extent to which they measure practices that affect student learning versus practices that affect customer satisfaction, the historical development of evaluation forms, and techniques that have been used to validate them.

Many of these issues can be illustrated by examining student ratings collected during the following case study, which involved several groups of professionals who attended lectures on the topic of game theory and physician education. A paraphrased account of these lectures, as described by Natfulin, Ware, and Donnelly (NWD), follows.

Eleven professional psychiatrists, psychologists, and social workers sat attentively through a lecture entitled "Mathematical Game Theory as Applied to Physician Education." The lecturer, Dr. Myron L. Fox, an authority on the application of mathematics to human behavior, gave an enthusiastic lecture peppered with humorous anecdotes and witticisms. At the end of the session, the participants completed a questionnaire in which Dr. Fox's instructional skills were evaluated. Not surprisingly, a majority of the group rated Dr. Fox's performance positively: 90% of the participants felt that Dr. Fox was enthusiastic, well organized, and interesting. All agreed that Dr. Fox had stimulated their interest in the topic and seemed himself to be quite interested in the topic as well.

Given the success of this lecture, a videotape of the same lecture was later presented to another group of mental health educators, psychiatrists, and psychologists. Their response to the lecture was similar to that of the first group; a majority of the group agreed that Dr. Fox was well organized, interesting, enthusiastic, made good use of examples, and had a deep interest in the subject at hand.

The videotaped lecture was then shown to an audience of 33 educators and administrators enrolled in a graduate-level university educational philosophy course. Twenty-one members of the audience held master's degrees, and eight others held bachelor's degrees. Like the previous groups, they were also impressed with Dr. Fox's lecture. Ninety-seven percent of the group thought that Fox was interested in his subject, 91% felt he used adequate examples in his explanations, 70% thought he was well organized, 87% felt he had stimulated their thinking on the topic, and 81% thought he presented the material in an interesting way [NWD73, 632–633].

In reality, however, Dr. Myron Fox did not hold an advanced degree in any subject and was not an expert in mathematics, game theory, or behavioral sciences. Instead, Dr. Fox was a professional actor who had been hired to present a content-free lecture filled with "double talk, neologisms, non sequiturs, and contradictory statements" [NWD73, 631].

The purpose of the Naftulin, Ware, and Donnelly study was to demonstrate that a sufficiently expressive and talented speaker could seduce students "into feeling satisfied that they have learned despite irrelevant, conflicting, and meaningless content" [NWD73, 630]. The publication of this ruse shocked many in the educational establishment, while confirming the suspicions of others already concerned over the proclivity of many instructors to entertain rather than educate students.

To counter the threat that the NWD study posed to the general acceptance of student evaluations of teaching as valid measures of teaching effectiveness, many educational psychologists responded by criticizing the design of the study. And in fact, there were legitimate concerns.

For one, the professionals who made up the audience in the NWD experiment were not subjected to a test after the lectures, and knew in advance that they wouldn't be. So it could be argued that they were less vested in the presentation than typical students are in classroom settings. Also, there was no control group in the NWD study, and hence no mechanism for assessing the impact of instructor expressiveness and lack of lecture content on student achievement. In addition, participant exposure to the ruse was unrealistically brief, lasting less than two hours. Whether this type of deception could be maintained over the course of an entire semester was not clear. Finally, the questionnaire itself was challenged on the grounds that the items presented for evaluation did not accurately measure the multiple dimensions involved in effective teaching (e.g., [CN75]).

To address several of these criticisms, Ware and Williams later repeated the original Fox study in a more elaborate design that included student outcome measures. In the revised study, approximately 200 undergraduate and graduate students taking a general studies course at Southern Illinois University were randomly assigned to one of six lecture groups. The six groups were defined according to whether the lecture content was high, medium, or low, and for each content level, whether the lecture was low in seduction or highly seductive. Content was defined according to the number of teaching points covered (26, 14, and 4, respectively), while seductiveness was defined according to the level of "enthusiasm, humor, friendliness, expressiveness, charisma, and personality" [WW75, 151] exhibited by the lecturer, who, as it turns out, was the same actor that had participated in the original NWD study.

After attending their assigned lecture, students were asked to complete an 18-item questionnaire that probed student opinions on various aspects of the lecturer's presentation, including the lecturer's knowledge, presentation style, enthusiasm, extent to which they felt they had learned covered material, etc. These 18 items were then summed to obtain a global score for the lecture. Students from each treatment group were also asked to complete a 26-item multiple-choice test that covered all 26 teaching points presented in the high-content lecture.

A statistical analysis of the test and lecture evaluation scores led Ware and Williams to conclude that although content affected test scores, it did not significantly affect students' evaluation of their lecture when the lecture format was highly seductive. That is, the satisfaction scores reported by students who received the highly seductive lecture were essentially the same regardless of whether students received the low-, medium-, or high-content lecture.[1]

For the low-seduction lectures, content was important in distinguishing between the scores reported for the high-content lectures and either the medium- or low-content lectures, but was not statistically significant in distinguishing between the medium- and low-content lectures. In addition, although lecture content was significant in predicting student achievement, it was important as a predictor of student satisfaction scores only for low-seduction lectures.

The possibility that increased exposure time would eliminate the Fox effect, as the effect of educational seduction had come to be known in the educational research community, was examined in still another study by Williams and Ware [WW77]. This experiment closely replicated the experimental conditions employed in [WW75] (see also [WW76]), with the exception that students were presented two lectures instead of one. Conclusions based on the extended exposure time were consistent with their earlier study and with NWD.

An interesting corollary study that examined the influence of student exposure times on course evaluations was later conducted by Ambady and Rosenthal [AR93]. In that study, a control group of college students was shown *three ten-second, muted* videotaped snippets extracted from lectures presented by graduate teaching fellows to a second group of undergraduates. The control group was then asked to complete a survey on which they rated the nonverbal behaviors of the graduate teaching fellows. The second

---

[1]Interestingly, evaluations from students who received the medium-content lecture were, on average, highest, although the difference with lectures of high and low content was not statistically significant.

group, consisting of students enrolled in a semester-long course with their teaching fellow, completed teacher–course evaluation forms at the end of the semester. Remarkably, the ratings of the teaching fellows' nonverbal behaviors collected from the first group of students—who had viewed only three ten-second voiceless, and thus content-free, videotaped segments of the teaching fellows' presentations—correlated highly with the evaluation scores of students who attended regular lectures with the same instructor over an entire semester. The correlation coefficients between the course evaluations returned by the regular students at the end of the semester and the evaluation scores assigned by the students who had seen only 30 seconds of randomly selected, voiceless videotape ranged from 0.26 to 0.84, with a mean of 0.61!

Meier and Feldhusen [MF79] provided further evidence of the dangers of educational seduction in yet another Fox-effect experiment. The major innovation of their study was the meticulous design of a questionnaire (the Microteaching Rating Scale, or MRS) to measure student opinion in several teaching dimensions. In addition to ascertaining student views concerning the amount of substantive content covered in a sequence of videotaped lectures, Meier and Feldhusen's survey instrument included a series of items intended to measure student opinions of instructor expressiveness, instructor personality, the ability of an instructor to explain course material, and instructor preparation. By including survey items to measure each of these aspects of teaching quality, the authors were able to examine the extent to which the effects of instructor expressiveness and personality could be disentangled from other instructor traits. Meier and Feldhusen also attempted to gauge the influence of the stated purpose of the evaluations by administering the questionnaire under two pretexts: Approximately one-half of the students were told that the questionnaire was to be used for administrative purposes, the other half were told that the questionnaire was to be used by the instructor for self-evaluation.

The results of Meier and Feldhusen's experiment support the earlier results reported by Ware and Williams and NWD:

The manipulation of lecture content (high vs. medium) significantly influenced responses to only one of the five student rating subscales. On the instructor explanations subscale, mean ratings of students who viewed the medium-content lectures were higher than the ratings for the high-content lectures. Responses on the lecture content subscale were *not* affected by the manipulation of lecture content [MF79, 343].

In regard to instructor expressiveness, Meier and Feldhusen reported that

> ... the expressiveness of the instructor had a significant effect on scores obtained from all five student rating subscales. Students who viewed the high-expressive lecturer gave higher ratings on the dimensions of instructor expressiveness, instructor personality, instructor explanations, instructor preparation, and lecture content than did students who viewed the low-expressive lecturer. In addition, the high-expressive lecturer received significantly higher ratings ($p < 0.05$) for 21 of the 23 items on the MRS than did the low-expressive lecturer [MF79, 342].

Meier and Feldhusen found that the stated purpose for the evaluation did not have a statistically significant effect on any of the five student rating subscale scores.

Additional educational seduction experiments continued in the years following the appearance of the Meier and Feldhusen study. Abrami, Leventhal, and Perry summarized the results from these experiments as follows:

> Thus, we can summarize the collective findings of educational seduction research in two statements. First, instructor expressiveness had a substantial impact on summary and global ratings but much less on student achievement. Second, lecture content had a substantial impact on student achievement but much less on student ratings [ALP82, 454].

In other words, lecture content is not important if you are a highly expressive, seductive lecturer; it doesn't affect student ratings.

SET Development

Despite the concerns raised by these studies over the validity of student evaluations of teaching, the educational establishment has been loath to eliminate teacher–course evaluations as the primary, and often only, mechanism for evaluating instructor proficiency. Administrators had at the time of these studies, as they do today, "a pressing need for valid and reliable instruments for measuring teaching behavior to support decisions intended to improve university teaching. Such decisions range from those of individual faculty members who desire to improve their teaching to those made by college administrators and outside agencies charged with the responsibility of rewarding meritorious performance and allocating scarce resources" [CN75, 430].

The appeal of student evaluations stems from the apparent fact that students have the greatest opportunity to observe instructors. Student evaluations can be obtained at essentially no cost, and the use of teacher–course evaluations obviates the need to establish more complicated, expensive, and controversial methods for assessing teaching quality. Student evaluations of teaching also provide a useful index of student (and alumni) satisfaction, which, of course, has serious implications in student–administration relations and, ultimately, in university fund-raising activities and undergraduate recruitment.

Student evaluations of teaching also serve as the primary tool available to educational psychologists for collecting data on classroom behaviors. Herbert Marsh, a leading advocate of and expert on teacher–course evaluations made this point quite clear when he stated that "research on teaching involves at least three major questions: How do teachers behave? Why do they behave as they do? and What are the effects of their behavior? ... Student ratings are important (in answering these questions) both as a process-description measure and as a product measure" [Mar84, 707]. Interestingly, the same researchers who champion the use of student evaluations of teaching as a tool for studying teacher behaviors seem to have little apprehension about the impact that their use has in determining these behaviors. Indeed, understanding the influence of student evaluations of teaching on teacher behavior may well be as important as understanding the extent to which student evaluations of teaching provide valid indicators of teaching performance.

The prevalence of student evaluations of teaching has virtually ensured that challenges to their validity would be either overcome, circumvented, or ignored. The approach most commonly taken in an effort to overcome validity questions has been to focus on the measurement of specific dimensions associated with effective teaching. By narrowing the scope of the questions that students were asked, researchers hoped to separate various aspects of teaching behavior from one another, and by so doing, obtain more valid measurements of each. In fact, even the concept of validity was viewed as problematic by many, and some argued that no widely accepted measure of student achievement could be defined to validate forms. Furthermore, factors that some considered biases were proposed as accurate reflections of teaching ability by others. For example, prior student interest is associated with higher student ratings, but does this necessarily imply a bias? That is, is it not possible that more interested students might learn more, making the association between ratings and prior interest valid?

In the end, even questions of validity have been diffused by offering a multidimensional interpretation to this quantity as well. *Construct validity* was the term deemed appropriate for discussing whether a particular instrument could actually predict student achievement or learning. *Convergent validity* became the popular alternative to construct validity, and required only that the statistical properties of an instrument remain more or less stable when applied in related settings. Importantly, convergent validity was explicitly divorced from all student outcome measures.

Numerous evaluation instruments have been constructed to measure the multiple dimensions of effective teaching. Prominent among the earlier forms are the Endeavor Instructional Rating Form [FLB75], the Michigan State Instructional Rating System [War73], and the Student's Evaluation of Educational Quality form (SEEQ) [Mar82]. Because the SEEQ provided the basis for collecting data for literally dozens of research articles and provided a prototype for many of the forms used to evaluate teaching today, and because of its meticulous design, it is useful to examine the process by which it was constructed and validated.

SEEQ development began with administration of several pilot surveys to students, faculty, and administrators at the University of California, Los Angeles. Each survey consisted of 50 to 75 items thought to be associated with instructional quality. According to Marsh, the pilot surveys were first administered to students, who

> ... in addition to making ratings, were asked to indicate the items they felt were most important in describing quality of teaching. Similarly, staff were asked to indicate the items they felt would provide them with the most useful feedback about their teaching. Students' open-ended comments were reviewed to determine if important aspects had been excluded. Factor analysis identified the dimensions underlying the student ratings, and the items that best measured each. Reliability coefficients were compiled for each of the evaluation items. Finally, after several revisions, four criteria were used to select items to be included on the UCLA version of the SEEQ. These were: (1) student ratings of item importance, (2) staff ratings of item usefulness, (3) factor analysis, and (4) item reliabilities [Mar82, 77].

It is important to note that at no point in the development of the SEEQ were measures of actual student learning or student achievement used to determine which items were included in the final survey instrument. Items and, by extension, teaching dimensions, were selected initially from the opinions of students and teachers about the traits they felt were important for effective teaching. Items that probed these traits were subsequently selected through a statistical technique called factor analysis.

Factor analysis is based on the idea that one or more unobservable traits, or factors, underlie the generation of multivariate data. For example, designers of standardized tests, like the SAT, implicitly assume that student aptitude can be accurately summarized by measuring essentially two factors: a mathematics ability and a verbal ability. In the context of measuring instructional quality, the corresponding assumption is that, say, three or four factors—perhaps an instructor's concern, enthusiasm, organization, and subject matter expertise—underlie student responses to all items on a course evaluation form.

Factor analysis of the items ultimately included on the SEEQ led Marsh to conclude that this evaluation form "measures nine distinct components of teaching effectiveness that have been identified in both student ratings and faculty self-evaluations of their own teaching." The nine components of teaching effectiveness purportedly measured on the SEEQ are learning/value, enthusiasm, organization, group interaction, individual rapport, breadth of coverage, examinations/grading, assignments, and workload/difficulty [Mar82, 77].

The procedures used in the development of the SEEQ and related multidimensional student evaluation of teaching instruments were intended to produce evaluation instruments that were not susceptible to Fox effects or other extraneous biases. By probing student opinions on, say, instructor enthusiasm, researchers hoped to isolate the effects of enthusiasm and expressiveness from student responses to items probing other aspects of teaching. To validate these instruments and to verify that they had accomplished this goal, researchers relied on basically two types of validation studies.

The first and most commonly employed study was the multi-section validity study. In this design, several sections of the same course are used to study average differences in student achievement as a function of differences in average student evaluations of teaching. To avoid biases in study conclusions that might otherwise result from uneven distributions of student abilities across sections, multisection validity studies usually require that students either be assigned randomly to sections or that they enroll in sections without knowing who the instructor will be. When random or blind assignment is not possible, pretest measures are sometimes used to account for section differences in student achievement, although this practice seems to be less common. Importantly, it is standard practice for each section to follow a common course outline, to use the same textbook, to have similar course objectives, and to share either all examinations or, more typically, the same final examination. "Support for the validity of SETs is demonstrated when the sections that evaluate the teaching as most effective near the end of the course are also the sections that perform best on standardized final examinations, and when plausible counter explanations are not viable" [MD92, 169].

The value of multisection studies in determining the validity of student evaluations of teaching has been rigorously debated since the mid-1980s. Marsh and colleagues (e.g., [MD92, Mar80]) have been the primary antagonists in this debate, and have criticized multisection validity studies on several grounds. Among these are the small number of sections typically available in any study (often fewer than ten), difficulties in comparing findings across multiple studies (as attempted in, for example, [Coh80]), and practical difficulties encountered in randomization.

Another problem with multisection validity studies involves the standardization of courses taught by participating instructors. In most studies, critical decisions regarding the scope of material covered, the pace of the course, the level of lectures, the difficulty of exams and homework assignments, and grading policies, along with many other aspects of course design, are wrested from the instructor and replaced by parameters determined by the study coordinator. In practice, however, these decisions determine precisely those course attributes that are most important in defining course content and student learning. Excluding these factors from the experimental design all but precludes an examination of their effects on student learning, and arguably reduces the scope of multisection validity studies to an examination of factors relating solely to instructor presentation style and personality.

Ironically, results from multisection validity studies are often cited as evidence that grading practices have little or no effect on student evaluations of teaching. But students' grades in these studies are normally determined, at least in part, by common tests, final exams, and homework assignments. In many cases, these assignments are even graded by the same teaching assistants. It is thus highly improbable that grading policies differ significantly between sections. How, then, can these studies be used to evaluate the impact of differences in instructor grading practices on student evaluations of teaching?

Perhaps the most disturbing aspect of multisection validity studies is the minuscule proportion of variance in student achievement that is normally accounted for by student evaluations of teaching—usually less than 20% [Coh87, KM95]. This fact led Marsh and Dunkin to state that "most variance in achievement scores at all levels of education is attributable to student presage variables and researchers are generally unable to find appreciable effects due to differences in teacher, school practice, or teaching method" [MD92, 170]. This concern is echoed even by supporters of multisection validity studies, including Abrami, D'Appollonia, and Rosenfield, who opine that "instructors may have genuinely small effects on what students learn" [AdR97, 170]. If this is, in fact, the case, the real issue may not be whether multisection validity studies can be used to validate teacher–course evaluation forms, but may instead be the very premise that the specific teaching behaviors "measured" on these forms are important to student learning at all.

The second approach taken toward establishing the validity of student evaluations of teaching focuses on convergent validity. Whereas construct validity requires that responses on an evaluation instrument correlate with an agreed-upon outcome measure (i.e., student learning), convergent validity requires only that the properties of an evaluation instrument remain stable across different settings. For example, to demonstrate the convergent validity of the SEEQ, Marsh and Dunkin [MD92] collected SEEQ rating data under varying scenarios, including ratings of the same instructor teaching two distinct courses, ratings of the same instructor teaching the same course on different occasions, ratings of two instructors teaching the same course, and ratings of different instructors teaching different courses. In this analysis of the SEEQ, the estimated correlation of overall instructor rating for instructors teaching the same course twice was 0.72, for the same instructor teaching different courses 0.61, and for different instructors teaching the same course −0.05. The estimated correlation of overall instructor rating for different instructors teaching different courses was −0.06. Marsh and Dunkin interpreted these results as providing evidence for the convergent validity of overall instructor rating, since the correlation of ratings received by the same instructor in different courses was nearly as high as the correlation of between-instructor ratings for the same instructor teaching the same course. Furthermore, the correlation of ratings of different instructors teaching the same course was essentially zero, providing further evidence that the course effect on this rating item was negligible.

The convergent validity of items designed to probe specific teaching dimensions exhibited similar behavior. In the same study, Marsh and Dunkin found that correlations between ratings of the same instructor teaching the same course ranged from .63 to .73 for the teaching dimensions measured on the SEEQ, while the corresponding correlations between ratings of the same instructor teaching different courses ranged from .40 to .61. The lowest correlations in the latter cases occurred for those items that the authors considered most course-dependent ("assignments" and "workload/difficulty").

Convergent validity of student ratings can also be investigated by comparing instructor self-evaluations with student evaluations. For the SEEQ, the correlation of instructor and student evaluations on similar teaching dimensions ranges from approximately 0.17 to 0.69 [Mar84]. However, conclusions based on such analyses must be tempered, because, as Centra [Cen79] points out, instructors may calibrate self-evaluations based on their previous student evaluations.

Other comparisons useful for examining convergent validity of student ratings include correlations of peer ratings and student ratings, and peer ratings and self-evaluations. Findings of Morsh, Burgess, and Smith [MBS56] and Centra [Cen75], however, suggest that the value of peer ratings of instruction are only weakly associated with both student achievement and student evaluations of teaching.

Overall, there seems to be little dissent in the educational research community concerning the convergent validity of well-designed student evaluation of teaching instruments. Abrami, D'Apollonia, and Rosenfeld summarize sentiment on this issue when they state that "the reliability of student ratings is not a contested issue: The stability of ratings over time and the consistency of ratings over students (especially in classes of ten or more) compares favorably with the best objective tests" [AdR97, 224].[2]

Still, the question remains as to whether the convergent validity of well-designed teacher–course evaluation forms is at all relevant in assessing the usefulness of these forms for predicting student performance and learning.

To illustrate this concern, suppose an item that probed student opinion of an instructor's hair color was inserted on a teacher–course evaluation form. This item would undoubtedly exhibit a high degree of convergent validity: Students in different sections would tend to agree on the instructor's hair color, as would the instructor's peers and even the instructor himself, and these views would be (relatively) stable over time. Indeed, this item would probably have higher convergent validity than nearly every other item on any chosen form. Furthermore, this item would have high divergent validity: Students in different sections would probably agree that different instructors had different hair color. But despite the wealth of favorable statistical properties that this item might possess, instructor hair color probably doesn't provide much information regarding the extent to which an instructor facilitated learning.

In short, the fundamental problem that arises in using measures of convergent validity to validate student evaluations of teaching is that these measures rely entirely on the largely unsubstantiated supposition that the items included on the forms can be reliably associated with improved student learning. They often can't.

Where, then, do we stand in our quest for meaningful student evaluations of teaching a quarter of a century after the publication of the original Dr. Fox experiments? The answer to this question is best provided in an educational seduction experiment conducted by Williams and Ceci [WC97] in 1997.

Their experiment began when one of the authors, Professor Ceci, was invited by his university to attend a teaching skills work-

---

[2]That is not to say, however, that researchers necessarily agree on which forms best measure teaching effectiveness or the factors that underlie it. For example, the SEEQ form and its creator Marsh identify nine dimensions of effective teaching; Feldman [Fel76] identifies 19 factors; Feldman [Fel88] lists 22 factors; and Kulik and McKeachie [KM75] identify four factors common to a meta-analysis of 11 other studies.

shop at the end of a fall semester. Ceci accepted the invitation, and together with Williams, decided to use the workshop as part of a "naturalistic experiment" to assess the effects of specific teaching skills on student evaluations of teaching. To this end, Ceci collected course evaluation forms from his students at the end of the fall semester for a developmental psychology course that he had taught for nearly twenty years.

The teaching skills workshop was conducted by a professional media consultant who was not a psychologist, and the workshop itself was attended by faculty from several departments. As a consequence, the workshop was not subject-specific but instead focused on presentation styles. Particular attention was paid to the use of hand gestures and techniques for varying voice pitch. Participants were also critiqued on a videotaped sample lecture.

During the following spring semester, Ceci initiated a number of steps to ensure that the conduct of the spring course was as close as possible to that of fall's. In particular, Williams and Ceci report the following controls:

> Specifically: (1) the same syllabus, textbook, and reserve readings were used each semester; (2) the same overhead transparencies were used at the same relative points each semester; (3) the same teaching aids (slides, videos, demonstrations) were used at the same points in each semester; (4) the identical exams and quizzes were given in each semester (with appropriate safeguards against their being removed from testing rooms by students); (5) nearly identical lectures were given in both semesters; (6) the course met on the same days of the week, at the same time, and in the same room both semesters; and finally, (7) the course had the same ratio of teaching assistants to students each semester.

> The point about the lectures being nearly identical in both semesters requires amplification. During the first semester, each lecture was audio recorded to preserve its contents (as was usually done by the professor for the benefit of students who miss class). Before the professor gave the same lecture in spring, he attempted to memorize what he had said in that lecture the prior fall. This was not as difficult as it might seem for two reasons. First, the professor had been teaching

this course since 1977, and on several occasions has taught it two to three times per year, thus resulting in a highly rehearsed set of lectures. Second, each lecture is accompanied by overhead transparencies that provide a detailed outline for that lecture. It is difficult to *avoid* repeating all of the substantive points made during the previous semester, inasmuch as the professor and class follow these outlines point by point [WC97, 16].

To verify the similarity of lectures after the completion of the spring semester, two raters unfamiliar with the purpose of the study were asked to compare eight randomly selected matched lecture pairs from the spring and fall to compare the substantive points covered in each. Williams and Ceci report 100% overlap of the lectures in this regard; every point covered in the fall lecture was covered in the spring, and vice versa. The only difference between Professor Ceci's instruction in the fall and spring semesters was his higher level of expressiveness in the spring.

In the final week of the spring semester, students were asked, as before, to complete a ten-item questionnaire, where each item was scored on a five-point scale with higher values indicating more favorable responses. Five questions probed instructor qualities, including the instructor's knowledge, tolerance to other points of view, enthusiasm, accessibility, and organization. The remaining questions concerned student views of how much they had learned in the course, whether goals were clearly stated, fairness of grading, text quality, and an overall course rating.

Not surprisingly, differences between student ratings of instructor enthusiasm in the fall and spring were highly statistically significant, increasing from a mean rating of 2.14 in the fall to a 4.21 in the spring. What is perhaps more surprising is the fact that highly significant differences ($p < .0001$) were also observed for every other item as well. Instructor knowledge ratings increased from 3.61 to 4.05; tolerance to others' views from 2.55 to 3.48; accessibility from 2.99 to 4.06; and organization from 3.18 to 4.09. Course ratings fared similarly: Amount learned increased from a mean of 2.93 to 4.05; clarity of goals from 3.22 to 4.00; grading fairness from 3.03 to 3.72; and overall course rating from 2.50 to 3.91. Even student ratings of the identical textbook increased from 2.06 to 2.98!

A critic of this study might argue that increased expressiveness of the instructor generated increased student interest in the subject matter of the course, and as a consequence may also have increased student attentiveness and learning. If so, higher student ratings would be justified. In fact, however, Williams and Ceci report that exam scores in the fall and spring were nearly identical. There was no increase in student learning.

Williams and Ceci's study also provides little support for the use of "divergent validity" measures for assessing multitrait evaluation forms. In this case, expressiveness significantly influenced all categories of student response, not just the item concerning enthusiasm.

However, the most telling summary of the Williams and Ceci study, and perhaps on the validity of student evaluations of teaching in general, is contained in D'Apollonia and Abrami's scathing commentary on Williams and Ceci's article, in which they claim that Williams and Ceci have contributed nothing new to the literature on student evaluations of teaching. In their accompanying discussion, D'Apollonia and Abrami make the following assertion:

> Moreover, what was learned by the present authors matches well with findings from prior research. As evidence of that, we computed the proportion of rating variance accounted for by expressiveness in the Williams and Ceci study. For mean instructor ratings, $\omega^2 = .27$. For mean course ratings, $\omega^2 = .21$. These are remarkably similar to the results reported in the 1982 Abrami et al. [ALP82] research integration [WC97, 18].

Thus, the magnitude of the effect of expressiveness reported by Williams and Ceci is judged by experts in the field of instructional evaluation to be consistent with the magnitude of the effect found in a large number of previous studies. That is, D'Apollonia and Abrami do not consider the results of Williams and Ceci's study surprising. Instead, they claim that the effect of expressiveness noted in this study is typical of its effect on most other teacher–course evaluation forms. But this, of course, is scandalous: The effect that Williams and Ceci report is enormous. If it is typical of the effect that expressiveness has on other forms, should these forms really be used for faculty assessment?

TOWARD PRODUCT MEASURES OF STUDENT ACHIEVEMENT

M any of the problems associated with student evaluations of teaching stem from ambiguities in their intended purpose. From the 1970s forward, educational researchers strove to design forms that were suitable primarily as aids to instructors who wished to improve their teaching techniques. For this purpose, the multidimensional approach to the design of teacher evaluation forms has been an unqualified success. Evaluation items that probe students' opinions on global course attributes, like "how much did you learn in this class," or "how did this instructor compare to others," clearly provide little guidance to instructors on how to improve their teaching when their ratings on such items are low. Furthermore, researchers have determined that feedback on ratings of specific teaching dimensions is associated with subsequent improvements on those teaching dimensions (e.g., [MR93]).

Despite these successes, the use of student evaluations of teaching for administrative purposes and as measures of overall teaching effectiveness has been an unqualified failure. Not only has their use for these purposes had the unintended consequence of altering the dynamics of student–instructor interactions in ways that are not yet fully understood, but current teacher–course evaluation forms are, at best, only modestly correlated with actual student achievement.

The use of multidimensional evaluation forms for administrative purposes is complicated further by the fact that administrative and faculty committees are usually not trained to interpret summary measures from factor analyses, and are not provided with guidelines as to how the various factor scores and item means should be weighted. As McKeachie points out, "such a committee must arrive at a single judgment of overall teaching effectiveness. If one grants that overall ratings of teaching effectiveness are based on a number of factors, should a score representing a weighted summary of the factors be represented (as Marsh and Roche [1997] argue), or should one simply use results of one or more overall ratings of teaching effectiveness (as contended by d'Apollonia and Abrami, 1997)?" [McK97, 1218]. Implementing Marsh and Roche's scheme requires determining an "optimal" weighting of factors, which, in a slightly different guise, is the topic of this section. And as discussed previously, the use of items probing overall teaching effectiveness, as proposed by d'Apollonia and Abrami, subjects faculty members to personnel decisions that may be heavily influenced by biases arising from the interplay of nonteaching behaviors (e.g., grading practices) with student assessments of more relevant instructional characteristics.

The failure of teacher–course evaluations to accurately predict student achievement can be traced to the methods used in their design. Like the development of the SEEQ instrument, most evaluation forms are created in student/faculty committees charged with identifying instructional processes judged to be important for effective teaching. Generally speaking, items selected by these committees are not subsequently calibrated with actual measures of student learning. On better forms, convergent validity of items or underlying factors is examined to refine the forms for the purpose of improving their reliability, but item pools are seldom, if ever, based on their construct validity.

One reason that items on teacher–course evaluations are not more often tied to measures of student achievement is that there are no universally agreed-upon measures of student achievement. For instance, Abrami, d'Apollonia, and Rosenfield define the products of effective teaching as the

> ... positive changes produced in students in relevant academic domains including cognitive, affective, and occasionally the psychomotor ones .... Included in the cognitive domain are both specific cognitive skills (e.g., subject matter expertise), general cognitive skills (e.g., analytical thinking), and meta-cognitive skills (e.g., error correction). Included in the affective domain are attitudes and interests toward the subject matter in particular and learning in general as well as interpersonal skills and abilities relevant to learning and working in a social context. Finally, included in the psychomotor domain are physical skills and abilities ranging from those acquired in a physical education to precise motor skills acquired in a fine arts education" [AdR97, 217].

Given the complexity of this product definition of teaching effectiveness, it is perhaps not surprising that educational researchers have opted instead for the process definition of teaching effectiveness when they set out to validate an instrument! Nonetheless, if student evaluations of teaching are to be used for tenure and promotion decisions, a product definition must be given precedence, even if it is difficult to agree on how student achievement and learning are to be gauged.

How, then, can effective teaching be measured? Use of standardized examinations is not feasible at the university level due to the large number of courses taught, the different emphases that instructors give to the same course, and the fact that many courses are taught by only a single faculty member. Multisection validity studies permit concrete comparisons of student achievement across sections of the same course, but require an enormous amount of effort to implement and, as a result, are unlikely to be performed on a routine basis. Besides, they suffer from the effects of unwanted standardization, as discussed above.

A practical solution to this conundrum might be found by basing measures of student learning on performance in follow-on courses, rather than on standardized tests or end-of-term examinations. Under such a scheme, instructor effectiveness in, say, a first-semester language course would be judged by student success in the second-semester language course. Similarly, first- and second-semester courses in mathematics, natural sciences, social sciences, and the humanities would be used to investigate the impact of various teaching dimensions on subsequent student achievement in third- and fourth-semester courses.

Critics of such an approach might argue that it too narrowly defines the products of effective teaching. No emphasis is placed on the development of general cognitive and metacognitive skills, or on the affective attitudes of students toward the subject matter of the course, or the development of students' interpersonal skills and abilities to work well within a social context. Furthermore, the idea of measuring student achievement in a prerequisite course by student success in follow-on courses most readily applies to courses for which prerequisites are taught in multiple sections. The extension of findings from introductory courses to higher-level courses having multiple prerequisites, which are often taught in a single section, is not completely transparent.

Still, the simplest measure of teaching effectiveness in a first-semester calculus course is the preparation of students for second-semester calculus, and average student performance in intermediate Spanish is an obvious measure of the effectiveness of instruction in introductory Spanish. And unlike multisection validity studies that impose unnatural constraints on teachers and teaching processes, the use of prerequisite course attributes to determine the construct validity of teacher–course evaluation forms does not require experimental manipulations of students or courses. Instead, it can be achieved using existing data from most university registrars' offices.

To test the feasibility of this approach for establishing the construct validity of student evaluations of teaching, a preliminary study was conducted using data collected during the DUET experiment. Sixty-two courses taught in the spring semester of 1999 were identified as having prerequisite course requirements that had been taught in multiple sections the preceding fall. The particular courses chosen for this study are listed as an appendix to this chapter and include courses in all major academic divisions.[3]

Using the course sequences listed in the appendix, average responses to the 26 DUET items that probed specific teaching behaviors were computed for each of the prerequisite courses. These variables, along with the average grade in the prerequisite course, the students' grades in the prerequisite course, and each student's GPA (computed without the grade received in the follow-on course) were used as covariates in an ordinal regression model designed to predict the grades received by each student in their follow-on course. In other words, the probability that a student received a particular grade in the advanced course was regressed on prerequisite course attributes, students' GPAs, students' grades in the prerequisite course, and the average grade in the prerequisite class.

---

[3]Unfortunately, the limited amount of data available for this analysis required the inclusion of several course sequences in which prerequisite courses were only tangentially related to follow-on courses. It should therefore be noted at the outset that including such sequences clearly diminished the sensitivity of analyses in detecting those course attributes that play an important role in predicting student achievement. A more comprehensive database with more stringently defined course sequences would certainly yield more definitive results. Furthermore, language courses appear in frequencies that are disproportionate to their enrollments; their effects on study conclusions are discussed in the sequel.

FIGURE 1

Illustration of ordinal data model for success in follow-on courses. The middle of the density is denoted by M. In this figure, the probability that a student receives a particular grade is given by the area under the density in the corresponding category. As the center of the density M moves to the right, the probability that a student receives a higher grade also increases.

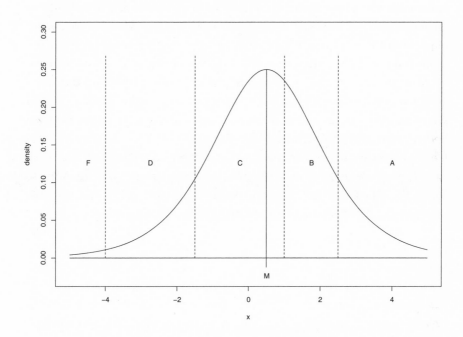

Although a brief description of regression analysis was provided in the last chapter, a simplified overview of ordinal regression is perhaps worth repeating here. In this context, the essential idea is that the probability that a student receives a particular grade in a course varies in a systematic way as a function of underlying covariates. Heuristically, this dependence is depicted in Figure 1. In this figure, the area under the density function within each category represents the probability that a student receives that grade. Explanatory variables, or covariates, have the effect of moving the density function to the left or right. For example, the middle of the density, M, might be assumed to take the functional form

$$M = \beta_0 + \beta_1 \times \text{Enthusiasm}, \tag{5.1}$$

where "Enthusiasm" represents the average rating of instructor enthusiasm recorded for the relevant prerequisite course. In this case, the effect of a one-unit increase in the average rating of enthusiasm would be to move the density in Figure 1 β_1 units to the right. Once moved, the area under the curve in the higher grade categories increases, reflecting the fact that the model then assigns a higher probability to the event that a student who had an enthusiastic instructor would receive a higher grade in the follow-on course than would a student who had a less enthusiastic instructor.[4] Large values of β_1 would therefore be associated with a strong effect of instructor enthusiasm on student learning. Values near zero suggest that enthusiasm is not an important predictor of student learning, and negative values suggest that students learn more with instructors who are not enthusiastic.

Including average student responses on all 26 DUET course attribute items and the three grade variables into an equation similar to (5.1) resulted in a maximum likelihood estimate for the center of the logistic distribution depicted in Figure 1 of the form[5]

$$M = \beta_{\text{class}}$$
$$+ 1.53 \times \text{grade in prereq } (.11)$$
$$+ 3.83 \times \text{GPA } (.19)$$
$$- 1.13 \times \text{challenging classes } (.74)$$
$$+ 1.94 \times \text{relevant classes } (.57)$$
$$+ 1.86 \times \text{challenging assignments } (.81)$$
$$- 0.941 \times \text{relevant assignments } (.84)$$
$$+ 1.70 \times \text{knew goals } (2.3)$$

[4] In the following, the density function in Figure 1 is assumed to be a standard logistic density, so the particular ordinal regression models fitted were proportional odds models. To account for differences in grading patterns of instructors in follow-on classes, a separate intercept term was included in the model for each of the follow-on courses.

[5] For clarity, the signs of the coefficients in this equation have been written so that they represent the effect of the covariates on the central value of the logistic density, rather than the more conventional notation in which the logit of the cumulative probabilities is expressed as the linear predictor subtracted from a threshold value.

$$- 0.20 \times \text{hours in class} \ (.78)$$

$$+ 0.065 \times \text{hours on assignments} \ (.43)$$

$$- 0.39 \times \text{completion of reading} \ (.49)$$

$$+ 0.20 \times \text{completion of written} \ (.55)$$

$$+ 1.03 \times \text{class attendance} \ (.97)$$

$$- 1.04 \times \text{course difficulty} \ (.49)$$

$$- 0.57 \times \text{encouraged questions} \ (.57)$$

$$+ 0.33 \times \text{accuracy of exams} \ (.76)$$

$$+ 0.61 \times \text{grading stringency} \ (.63)$$

$$+ 0.17 \times \text{exam usefulness} \ (.52)$$

$$- 0.66 \times \text{instructor knowledge} \ (.44)$$

$$- 0.15 \times \text{instructor availability} \ (.66)$$

$$- 1.09 \times \text{instructor organization} \ (.53)$$

$$+ 0.66 \times \text{related research} \ (.41)$$

$$- 0.31 \times \text{critical thinking} \ (.38)$$

$$+ 0.19 \times \text{instructor concern} \ (.65)$$

$$+ 0.72 \times \text{instructor enthusiasm} \ (.55)$$

$$- 0.026 \times \text{instructor communication} \ (.73)$$

$$- 1.33 \times \text{mean prereq grade} \ (.70). \tag{5.2}$$

In this equation, standard errors of the regression coefficient estimates are displayed in parentheses after each term.

To understand how the numbers appearing in (5.2) should be interpreted, consider the second term in the equation, $+1.53 \times$ "grade in prereq." The coefficient $+1.53$ means that the center of the curve in Figure 1 shifts 1.53 units to the right for every one unit increase in the grade that the student received in the prerequisite course. In other words, a student who got an A has a curve that is 1.53 units farther to the right than a student who got a B, and so has a correspondingly higher probability of getting a higher mark in the follow-on course. Similarly, the penultimate term, corresponding to instructor communication, implies that a one-unit increase in the average student response to the DUET item that probed the extent to which the instructor was able to effectively communicate course material (item 34) in the prerequisite course had the effect of moving the logistic density depicted in Figure 1 to the *left* 0.026 units.

The standard errors in (5.2) indicate how well the coefficients were determined using available data. In general, a 90% confidence interval for each coefficient can be obtained by adding and subtracting 1.64 times the value of the standard error to the value of the coefficient. So, for example, a 90% confidence interval for the effect of instructor communication in determining the grade of a student in the follow-on course can be calculated as ranging from −1.46 to 0.94. This range, from −1.46 to 0.94, is relatively wide when compared to the coefficient value −0.26 and includes the value of 0. Evidently, it was not possible even to determine whether the effect of instructor communication was positive or negative using just these data, which consisted of the grades of 1,263 students received in 62 follow-on classes requiring one of 58 prerequisite courses.

The fact that the 90% confidence interval for the effect of instructor communication includes 0 might loosely be interpreted as meaning that this variable was not statistically significant at the $1 − .90 = 0.10$ level, or the 10% level. This result, that instructor communication was not statistically significant as a predictor of student success in the follow-on class, could be the result of two factors. First, it might mean that differences in instructor communication just didn't have much effect on student learning, at least within the range of instructor communication skills possessed by Duke teachers. That is, if all Duke teachers were highly effective communicators, this item wouldn't be relevant for predicting student success in follow-on courses. Second, the insignificance of the item probing instructor communication might mean that more data are needed to estimate the size of the effect. Or it could be a combination of the two. What is clear, however, is that for these data, the effect of instructor communication was not strong enough to make it distinguishable from random variation in student achievement across semesters.

Perhaps the most striking aspects of equation (5.2) are the relatively small coefficients associated with many of the other teaching behaviors, particularly when these coefficients are compared to their standard errors. Instructor communication is but one example. In fact, only 4 of the 26 coefficients of the teaching behaviors were statistically significant at the 10% level. The significant explanatory variables included relevance of classes, the extent to which students found assignments challenging, course difficulty, and instructor organization. Interestingly, both student ratings of course difficulty and instructor organization were *negatively* associated with student performance in the follow-on course. Also, average grade in the prerequisite course was negatively associated with follow-on performance. The best predictor of student performance in the follow-on course was student GPA (computed without the contribution from the follow-on course).

A difficulty that arises in interpreting the regression coefficients in (5.2) involves the collinearity of many of the DUET items. This means that student responses to many DUET items are correlated with each other, and this has the statistical effect of inflating the standard errors of the coefficients for those items that are highly correlated with others. One way of circumventing this collinearity is to eliminate terms from the equation that are not statistically significant. By getting rid of variables that don't seem to make a difference on their own, the effects of the remaining variables can be isolated and more accurately pinned down. Eliminating unimportant terms from the regression function also has the advantage of making the equation easier to interpret.

With these considerations in mind, statistically insignificant terms in equation (5.2) were successively eliminated, beginning with the terms that had the smallest ratio of their coefficient to their standard error. This continued until the magnitudes of all remaining ratios were significant at the 10% level. That is, terms were eliminated until the values of the ratios of all remaining coefficients to their standard errors were either greater than 1.64 or less than -1.64. This resulted in the following reduced equation for predicting student success in the follow-on classes:

$$M = \beta_{\text{class}}$$

$$+ 1.52 \times \text{grade in prereq} \ (0.11)$$

$$+\,3.82 \times \text{GPA } (0.19)$$

$$-\,0.89 \times \text{challenging classes } (0.41)$$

$$+\,1.51 \times \text{relevant classes } (0.37)$$

$$+\,1.10 \times \text{challenging assignments } (0.34)$$

$$+\,0.62 \times \text{class attendance } (0.36)$$

$$-\,1.15 \times \text{course difficulty } (0.38)$$

$$-\,0.47 \times \text{encouraged questions } (0.26)$$

$$+\,0.79 \times \text{grading stringency } (0.35)$$

$$-\,0.61 \times \text{instructor knowledge } (0.29)$$

$$-\,0.71 \times \text{instructor organization } (0.29)$$

$$+\,0.58 \times \text{related research } (0.22)$$

$$+\,0.65 \times \text{instructor enthusiasm } (0.23)$$

$$-\,1.41 \times \text{mean prereq grade } (0.39) \tag{5.3}$$

The interpretation of coefficients in this equation is still difficult owing to the observational nature of the data acquisition process, persisting collinearity between covariates, and the relatively small number of classes available for parameter estimation. Had more data been available, undoubtedly many of the items eliminated from (5.2) would also have attained statistical significance, although the magnitude and importance of their effects is less certain. Nonetheless, equation (5.3) does suggest several disturbing and potentially important associations among those variables that were found to be statistically significant at the 10% level.

To begin, the most important predictor of success in the follow-on class, in both (5.2) and (5.3), is student GPA. Evidently, good students tend to do better than poor students.

More surprising, however, is the estimated effect of the average prerequisite course grade. The sign of this coefficient is negative, which suggests that prerequisite courses in which instructors grade more stringently are more effective in preparing students for advanced courses. This interpretation is supported by the positive coefficient of stringency of grading; both the item that probed stringency of grading practices and the average grade in the prerequisite course suggest that stringent grading is associated with higher levels of achievement in follow-on courses. This finding directly contradicts the teacher-effectiveness theory, and provides additional support for the grade-leniency and grade-attribution theories.

Of course, alternative interpretations of the negative association between average prerequisite course grade and student achievement in follow-on courses can be constructed. For example, one might contrast the coefficient of average prerequisite course grade to the coefficient of individual student grade in the prerequisite course. Because these two coefficients are nearly identical in magnitude, yet opposite in sign, it might be posited that it is the difference in a student's grade from the mean section grade that is important in predicting success in the follow-on course. When student grade in the prerequisite course is eliminated from the regression equation, however, the coefficients of both grading stringency and mean grade in the prerequisite course maintain the same signs, although their magnitudes decrease. Furthermore, while this interpretation is plausible, it implicitly requires the acknowledgment that grading standards in prerequisite courses vary, and that evaluation of student grades should therefore account for such differences. This last point is revisited in Chapter 7.

Consistent with findings of Fox-effect studies, enthusiasm was positively associated with student achievement in follow-on courses. This result makes sense. Enthusiastic instructors tend to maintain higher interest levels among their students, and so their students tend to learn more.

Equations (5.2) and (5.3) suggest that student ratings of course difficulty are negatively associated with student achievement. While this finding may be evident to many, it may come as something of a surprise to others. There seems to be a common perception among many professors, particularly among those in the sciences, that "harder" courses force students to learn more. However, the results of this analysis suggest that students know what they are talking about when they report that course material is difficult:

When they say they don't get it, they don't. Qualitatively similar associations have been reported elsewhere [Mar84], and if valid, suggest that simply increasing the difficulty level of a course is not likely to increase the amount of material learned. Similar comments apply also to the negative coefficient associated with the proportion of classes judged to be challenging.

In contrast to course difficulty ratings, the proportion of reading and writing assignments that students perceived as challenging was positively associated with follow-on course performance. It may therefore be important to distinguish between making lecture presentations challenging, and challenging students on projects and homework.

The negative associations between student achievement in follow-on courses and both student perceptions of instructors' knowledge of course material and instructors' classroom organization are somewhat counterintuitive. In the case of the negative association with instructor knowledge, several explanations for this result might be forwarded, including the possibilities that (1) this association represents a statistical artifact due either to random variation in student responses or inadequacies in the experimental design, (2) presentation of complex material by instructors serves more to confuse students than to enlighten them, even though it may lead to an impression among students that instructors are highly expert, or (3) ratings of instructor knowledge served as a surrogate for instructor age in this analysis, and that older instructors have more difficulty "connecting" with students.

Finally, it is worth noting that the regression coefficients reported in (5.2) and (5.3) are relatively robust with respect to the course sequences included in the analyses. For instance, the magnitude of most of the regression coefficients reported in these equations remained approximately the same when language courses were excluded from the analyses, as they did when natural science courses were excluded.

The regression functions specified in (5.2) and (5.3) are not, of course, intended to be the final word on which teaching behaviors and course attributes are most important for predicting student learning. Nor are the interpretations of the regression coefficients offered above. Instead, this analysis is intended to highlight the dangers inherent in accepting the construct validity of items on course evaluation forms based only on "expert" opinion. The unexpected signs of the coefficients of instructor organization and knowledge, along with the small and statistically insignificant coefficients associated with items that probed many of the teaching dimensions— including instructor concern, hours per week spent in class, knowledge of course goals, and effectiveness of exams—suggests that although these items may be important indicators of customer satisfaction, they may not have a significant impact on actual student learning. In addition, the finding that student learning is negatively associated with mean prerequisite course grade raises further concerns over the consequences of undergraduate grade inflation.

As a mental exercise, it is interesting to contrast the current situation in which professors are evaluated using course evaluation forms that have high convergent validity to a hypothetical environment in which faculty assessment was instead based on a formula similar to (5.3). Under the latter scenario, professors would be rewarded not only for being enthusiastic, but would also be rewarded for behaviors that were demonstrably associated with increases in student achievement. Based on this preliminary analysis, such behaviors might include making lectures more transparent and less structured, assigning more challenging homework, relating course material to current research in the field, and maintaining high academic standards through the implementation of stringent grading standards. It would be interesting to observe the impact that such an evaluation scheme would have on recent trends in grade inflation and decreasing student workloads.

The most important aspect of the analysis presented above is that it can be repeated, using more courses and with greater precision, at many colleges and universities throughout the United States. The only requirements for its replication are the availability of historical course evaluation summaries and access to student enrollment and grade records. At most colleges, this would require only a minimal amount of additional data collection and management, and the potential benefits are enormous. Not only would such a practice

offer the potential for designing and implementing teacher–course
evaluation forms that are high in construct validity, but in so doing
would encourage the adoption of instructional techniques that are
directly linked to higher levels of student achievement.

CONCLUSIONS

Since the publication of the Naftulin, Ware, and Donnelly
[NWD73] study, concern over the validity of student evaluations of teaching has permeated academia. As recently as 1997,
experiments in educational "seduction" have demonstrated very
substantial biases to student evaluations of teaching that can be
attributed to behaviors as simple as instructor expressiveness
[WC97]. Similar concerns persist over the effects of grading policies
on student evaluations of teaching. Although little data is available
to quantitatively assess the impact that these perceptions have on
faculty behavior, there is no lack of anecdotal evidence to support
the notion that many professors have modified their instruction so
as to improve their teaching evaluations, possibly without actually
improving their teaching (e.g., [Nea96]).

Educational researchers and administrators have attempted to
overcome these concerns through the development of increasingly
sophisticated evaluation forms designed to measure specific teaching
behaviors. From a psychometric standpoint, this endeavor has led
to the development of evaluation instruments with high convergent
validity. Undeniably, these forms represent a major advance over
earlier instruments when they are used for instructor self-evaluation
and self-improvement. However, tools for accurately evaluating
effective teaching and student achievement remain elusive.

Current evaluation instruments provide little (legitimate) guidance to administrators in promotion, tenure, and salary decisions.
Abrami and d'Apollonia [Ad90, Ad91, AdR97] have long argued
that administrative committees are unable to interpret the relative
importance of various teaching dimensions vis-à-vis teaching effectiveness. In addition, the analyses presented above suggest that
many items on these forms may actually be negatively associated

with student achievement. At the very least, items on these forms explain only a small proportion of the variation in student learning from one class to the next. In fact, global items that probe teaching effectiveness might provide better summaries of teaching performance than do summary scores derived from specific items contained on multitrait evaluation forms [McK97]. Unfortunately, the sensitivity of global measures of teaching effectiveness to biases like expressiveness, grading leniency, and other factors is, as discussed previously, unacceptably high.

APPENDIX

TABLE 1

Course sequences used in the first part of Chapter 5. BAA is an abbreviation for Biological Anthropology and Anatomy, EGR for General Engineering, and PPS for Public Policy Studies.

Follow-on	Prerequisite	Follow-on	Prerequisite
BAA 132	BAA 093	JPN 002	JPN 001
BME 101L	EE 062L	LAT 002	LAT 001
BME 101L	MTH 111	ME 101L	PHY 052L
CHM 152L	CHM 151L	MTH 032	MTH 031
ECO 149	ECO 002D	MTH 032L	MTH 031L
ECO 149	ECO 052D	MTH 104	MTH 103
ECO 154	ECO 002D	MTH 103	MTH 032
ECO 154	ECO 052D	MTH 103	MTH 032L
EE 062L	EE 061L	MTH 103	MTH 041
EE 064	EE 061L	MTH 103L	MTH 032L
EGR 075L	PHY 051L	MTH 111	MTH 103
ENG 144	ENG 143	PHY 052L	PHY 051L
FR 063	FR 002	PPS 110	ECO 052D
FR 002	FR 001	RUS 002	RUS 001
FR 076	FR 063	RUS 063	RUS 002
GEO 043S	GEO 041	RUS 066	RUS 002
GER 002	GER 001	RUS 064	RUS 063
GER 065	GER 002	RUS 067	RUS 066
GER 066	GER 065	SP 002	SP 001
IT 002	IT 001	SP 063	SP 012

Grades and Student Course Selection

Grading policies affect students in a variety of ways, most obviously in the realm of student assessment. However, the most serious influence that instructor grading practices exert on students may well be their effect on student course selection decisions.

Using records of the course mean grades that students examined during the DUET experiment in conjunction with records of the courses for which they later registered, this chapter examines the influence that grading practices have on student enrollments. The conclusion of this investigation is that the influence is substantial. For courses taken as electives, students who participated in the DUET experiment were twice as likely to choose courses graded at an A– average as they were courses graded at a B average. This conclusion applies to courses chosen from within all academic divisions and likely results in a 50% decrease in the number of elective courses taken by undergraduates in the natural sciences and mathematics.

I T IS DIFFICULT TO OVERSTATE THE IMPORTANCE OF
student selection decisions to the success of America's higher
educational system. Neither modern computing and labo-
ratory equipment nor state-of-the-art multimedia classroom environments can compensate for a student being in the wrong class
for the wrong reason. Although the right reasons for students to
choose their classes vary from one student to the next, criteria
related to students' academic interests, career goals, and personal
development should always weigh heavily. Too often, however, less-
noble criteria—like manipulation of one's GPA—overshadow these
more legitimate concerns.

Those who have taught in American colleges are almost certainly aware of this dilemma; for those who haven't, the following
anecdotes illustrate the magnitude of the problem.

During the debate over grading reform at Duke during the
1996–1997 academic year, I was invited to discuss the merits of
the achievement index proposal at several student gatherings. The
vice president for academic affairs of the Duke student government
was responsible for organizing most of these events, so it was not
surprising that one of the talks was scheduled at his fraternity. On
the evening of that particular talk, I arrived at his frat house only
to notice that the pledge class had been organized into a neat row of
coats and ties at the back of the common room, and that each pledge
had been provided with a question that they were to raise at the end
of my presentation.

The questions asked by the pledges were well organized and
ranged over a variety of topics including the statistical properties
of the adjustment scheme, the impact it would have on students
applying for graduate and professional schools, and whether entities
outside Duke would be able to understand the adjustment. The

most telling questions, however, involved the potential social impact of the adjustment. This topic was broached when a student asked whether it was true that making adjustments to student GPAs using the achievement index would have the effect of downweighting A's assigned by instructors who graded more leniently than others. When I responded that it would do exactly that, an avalanche of questions followed, and a consensus that the achievement index would thus irreparably damage the social atmosphere at Duke was quickly reached. Several students candidly asserted that eliminating "gut" courses in which students could do a minimal amount of work and receive a decent grade would impose an unfair burden on students who had come to Duke for the overall college experience. Other students stated that such courses were necessary for students to balance academics and social activities, and that it was common practice for students to select at least one gut course each semester to balance their workload. Still others questioned whether it would be fair to require premed students to take three "real" courses when they were registered for organic chemistry, and whether this wouldn't hurt their chances of getting into medical school. And so on. Anyone present at this meeting would have come away with a clear understanding that a substantial portion of Duke undergraduates were aware of the differences between grading practices and workloads in various courses offered at the university, and that many students selected at least a subset of their courses based on these differences.

Faculty acceptance of this phenomenon is only slightly more subtle. In the fall of 2000, I was asked to attend orientation training after volunteering to become a premajor advisor. During one of the orientation classes, a fellow trainee asked whether it was possible for first-year students to enroll in more than five courses.[1] The facilitator thanked her for the question, and emphasized that it highlighted a more general point. Specifically, it was important that we, as premajor advisors, *strongly discourage* our first-year advisees from taking even four difficult courses in their first semester at Duke. Two case studies then followed. The story line in both was the

[1] The normal course load for an undergraduate at Duke is four courses.

same: A first-year student arrived at Duke with loads of advanced placement credit and an excellent academic record. Confident that he could handle chemistry, biology, and an upper-level calculus course, he took all three in his first semester. The student received a low grade in at least one of these courses and was thus effectively prevented from fulfilling his lifelong ambition of gaining admission to medical school. Or law school.

The message was clear. Encourage students to adopt the same strategy expounded by the pledge class: Lighten your load with adequate amounts of fluff.

As it turned out, the message was also unnecessary. Students who had been at Duke only two weeks already knew. During my first two interviews with premajor advisees, both asked how they could drop a science course and substitute a music class in its place. They had been advised by upperclassmen not to take too many hard courses in one semester. Not surprisingly, the particular music class they wanted to add was famous among Duke undergraduates for its lenient grading policies and light workload.

My purpose in recalling these episodes is not to advocate that students fill their schedules exclusively with strictly graded, time-consuming lab courses, or that students concentrate their schedules in the natural sciences and mathematics. Instead, my point is that students regularly enroll in courses because they wish either to reduce their workload or to obtain a high mark. While adding breadth and variety to one's background may be laudable, taking a course because it is an easy A is not.

The fact that college students engage in this type of behavior is, I think, widely accepted. Nonetheless, quantitative research to determine the extent of this problem is almost completely lacking. In one of the few studies that investigated factors that influence student course selection at the university level, Coleman and McKeachie [CM81] investigated the effects of course attributes on the selection decisions of prospective freshman political science majors in choosing between introductory political science courses. In their study, students were given course evaluation summaries for previous offerings of four first-year political science courses. Their course selection decisions were then examined as a function of the information that had been provided to them. According to the study's authors, students tended to choose "higher quality" courses even when workload requirements for these courses were higher. In contrast to results reported below, this study suggests that students prefer high workload courses when selecting from among courses in which they have a high interest level.

A more direct investigation of the influence of grades on student course selection decisions was performed a decade later by Sabot and Wakeman-Linn [SWL91]. Using transcript and survey data obtained from students at Williams College, the authors estimated the probability that students took a second course in each of five major departments as a function of their grade in their first course in that department. Also included in the probit model employed to estimate these probabilities were variables that represented students' intent to major in the given department, a gender variable, and a variable designed to measure a student's desire for achievement.

The five departments studied were Economics, English, Mathematics, Political Science, and Psychology. Statistically significant effects of students' grades in the first course were found to influence the probability that they took second courses in Economics, English, and Mathematics. In each case, the probability of taking a second course decreased with decreases in their grade in the first course. Similar effects were noted for Political Science and Psychology courses, although the estimated effects for those departments were not statistically significant.

In the department with the lowest mean grade, Economics, Sabot and Wakeman-Linn found that for male students who did not intend to major in economics, the probability of taking a second economics course was 18% lower for students who received a B in their first course than for students who received an A. The decrease in the probability associated with receiving a C rather than an A was 28%. For English courses, the probability that a male student not intending to major in English took a second course in English after receiving a B in his first course was estimated to be 14% lower than if he received an A, and 20% lower if he received a C. Furthermore, based on simulation studies using these probabilities, the authors estimated that the number of students taking one or more courses beyond the introductory level in Economics would increase by 12% if Economics employed the same grading practices as English; that the number of students taking one or more English classes beyond the introductory level would decrease by 14% if English used the same grading policies as Economics, and that if Mathematics employed the same grade distribution as an introductory English class at Williams, there would be an 80% increase in the number of Williams students taking at least one additional mathematics class. Finally, if English classes were graded using Mathematics' grade distribution, the authors predicted that the number of students taking a second course in English would decline by 47% [SWL91, 165–6].

Although the Sabot and Wakeman-Linn study does not examine factors related to students' initial decisions to take even one course in a department or the effects of anticipated grading practices in follow-on courses on students' subsequent enrollment decisions, it does provide valuable corroboration of the analyses presented below. Their analysis also seems to generalize beyond classes taken at Williams College in the late 1980s. Similar results apply also to Duke under-graduates, as suggested by the data depicted in Figure 1.

ANALYSIS OF THE DUET COURSE SELECTION DATA

When completing the DUET survey, students were asked to answer two questions concerning the effects that grading policies had had on their decisions to take the courses they had ultimately decided to take. The first, item 23, was phrased "On a scale of 1–5, with 1 being 'completely unaware' and 5 being 'completely aware,' how aware were you of how this course would be graded when you enrolled in it?" The second, item 24, was posed as "How much did your knowledge of how the course would be graded positively affect your decision to enroll in it?" Possible responses were "No effect or negative effect," "Slight effect," "Mod-

FIGURE 1

Replication of Sabot and Wakeman-Linn study for Duke University under-graduates. The three lines in this plot represent the estimated probabilities that students took a second course in a department as a function of their grade in their first course, for courses taken during their first semester at Duke University. For clarity, probabilities have been aggregated by academic division. Results displayed in this plot represent empirically observed probabilities and have not been adjusted for gender, intent of students to major in the corresponding departments, students' need for achievement, and whether or not the first course taken was used to (partially) satisfy a distribution requirement.

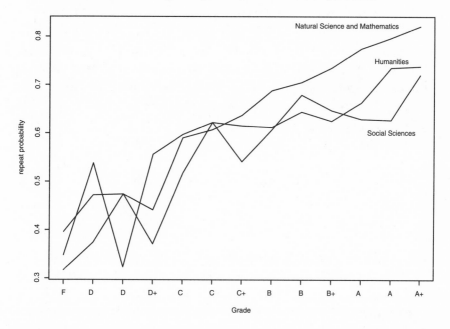

erate effect," "Significant effect," "Very significant effect," or "Not Applicable."

The survey also included a minor deception. After completing the form for the courses that they were taking (first-semester first-year students) or that they had taken in the previous semester, students were invited to examine both summaries of course evaluation data entered by other students or mean course grades recorded for courses taught in past semesters. What students didn't know was that a record was made of each course summary or mean grade that they examined. Over the course of the experiment, participating students viewed 9,434 histogram summaries of course evaluations and 42,325 mean course grades.

These data, along with students' responses to items 23 and 24, were used to assess the influence that grades had on student enrollment decisions by comparing the courses for which students had earlier viewed mean grades to the courses that they subsequently took. These comparisons were then examined in light of students' responses to items 23 and 24.

Several analyses were based on such comparisons. In the first, the effect of an instructor's grading policy on the probability that a student chose the instructor was investigated. To estimate how this probability varied as a function of the previous mean grades awarded by each instructor, a list of courses for which mean grades were viewed by the student in the fall of 1998 was compiled. This list was then compared to the list of courses that the student took in the spring of 1999. If the same course was identified in both lists, then the mean grade for that course, along with the mean grades of all other instances of the same course viewed by the student, were collected and saved for further analysis.

The following example illustrates this procedure. Suppose that student i, after completing the DUET survey in the fall of 1998, examined the mean course grades for three introductory calculus sections. Suppose that Section 1, taught by Instructor X in the spring of 1997, had a mean grade of 3.2. Section 2, taught by Instructor Y in the fall of 1996, had a mean grade of 3.4, and Section 3, taught by Instructor Z in the spring of 1997, had a mean grade of 3.1. If student i took introductory calculus in the spring of 1999 from one of instructors X, Y, or Z, then the mean grades (3.2, 3.4, 3.1) for the three sections of calculus, along with the particular instructor that the student chose, were recorded. For later reference, let the set of courses that student i considered taking be denoted by S_i, and let I denote the course that the student actually took. Also, let g_j denote the mean grade of the jth course the student considered taking. In this example, g_1, the grade in the first calculus course, was 3.2. If student i took the calculus course whose instructor previously had a 3.4 mean grade, then I would be 2, and S_i would simply refer to the set of the three calculus courses the student considered.[2]

The same procedure was repeated for the DUET grade-viewing records that were collected in the spring of 1999 and student registration data obtained from the university registrar's office in the fall of 1999. Combining the results over both semesters yielded a total of 197 courses for which students had both registered and viewed the mean grade of a previous offering of the same course with the same instructor, and for which the student had also examined the mean grade of at least one other version of the same course with a different instructor.

Considering again the example of the student choosing among three calculus courses, there is no reason, a priori, to think that any one calculus section would have a higher chance of being selected than any of the others. However, if grading practices did influence the student's enrollment decision, then instructors who had assigned higher mean grades in past courses would tend to be preferentially selected. The extent to which students did tend to take courses from such instructors thus provides evidence that students differentially selected courses with instructors who grade leniently. A convenient model for quantifying these tendencies arises as the natural generalization of the standard logistic regression model for multinomial data. Under this model, the probability that student i selects course j, given that student i selected a course from selection set S_i, is assumed to take the form

$$\frac{\exp(\beta g_I)}{\sum_{j \in S_i} \exp(\beta g_j)}, \tag{6.1}$$

for some unknown regression parameter β.

Although (6.1) might appear a bit daunting, its interpretation is actually quite straightforward. For example, if $\beta = 0$, then $\exp(\beta g_j)$ is 1 no matter what the value of g_j is, and so in that case (6.1) reduces to 1 divided by the number of courses for which the student had

---

[2]In the event that a student viewed the mean grade of several versions of the same course taught by the same instructor, all versions of the course were treated as being identical, and the mean grades for each replication of the course/instructor combination were averaged.

examined mean grades. That is, when $\beta = 0$ all courses are assigned equal probability of being selected. When $\beta > 0$, courses with high grades are assigned a higher probability of being selected; when $\beta < 0$ courses with low mean grades are assigned a higher probability of being selected.

For the data collected to estimate the effect of course grades on student selections of instructors, the maximum likelihood estimate of β was 0.94. Substantively, the importance of this value can be illustrated by considering a student who has the choice of taking one of two courses taught by two different instructors. If the mean grade in a previous offering of the course taught by the first instructor was 3.7 (i.e., an A— average) and the mean grade in a previous offering of the course taught be the second instructor was 3.0 (i.e., a B average), then the probability that a student would take the first course is estimated by

$$\frac{\exp(0.94 \times 3.7)}{\exp(0.94 \times 3.7) + \exp(0.94 \times 3.0)} = 0.66.$$

Similarly, the probability that the student would take the second course is 0.34. Thus, the odds that a student would choose the first instructor's course over the second's course were estimated as being nearly 2 to 1.

These results apply to students selecting from among different instructors of the same course. The effect of students choosing courses from among those courses that satisfy the same general educational requirements are examined next.

Prior to the fall semester of 2000, students matriculating at Duke University were required to take at least two courses in each of five of six academic areas (Arts and Literature, Civilizations, Foreign Languages, Natural Sciences, Social Sciences, and Quantitative Reasoning), and a third course, one of which had to be at the "100-level,"[3] in four of these five. To estimate the effects of grades on student course selection for students selecting from among possible

[3]Ostensibly, 100-level courses designate courses that have prerequisites and so are not considered introductory in nature. However, in addition to grade inflation, Duke also suffered from course inflation during the 1990s. By the late 1990s, many introductory courses at the university had been (re)designated as 100-level courses in order to reverse declining enrollments.

courses taken to satisfy a distributional requirement, the preceding analysis was repeated for courses selected within each of these academic areas.

More specifically, the list of classes for which each student viewed mean course grades was compared to the list of classes that the student subsequently took. If a class was common to both lists, and if the class satisfied a distributional requirement, the set of classes for which the student had viewed mean grades and that also satisfied the same distributional requirement was recorded. For example, if a student examined mean grades for previous offerings of STA 110A (introductory statistics), MTH 31 (first-semester calculus), and CPS 6 (introduction to computer programming), all of which satisfy the quantitative reasoning requirement, and then took any one of these courses, the selection set (S_i) would consist of all three courses. Instructors were not used to distinguish between courses in this analysis, and multiple examinations of the same course with or without different instructors were considered as a single course with a mean grade defined by averaging across the previous offerings. As before, the relationship between the mean grades of the courses in the selection set and the probability that a student selected a particular course was assumed to take the form (6.1). Table 1 lists the maximum likelihood estimates of β obtained by considering courses taken to satisfy each distributional requirement.

The regression coefficients for Civilizations, Social Sciences, and Arts and Literature all take similar values, and are also similar to the value obtained by defining selection sets on the basis of instructors within the same course. The large standard error associated with the regression coefficient in the Foreign Language category makes its value difficult to interpret and is a consequence of the fact that only four language course selections were captured in the selection procedure.

The negative value of the Natural Sciences' coefficient contrasts sharply with the positive coefficients estimated for the other distributional requirements. Though not statistically significant, its direction suggests that students differentially selected natural science courses with *low* mean grades.

TABLE 1

Maximum likelihood estimates of the course selection parameter β for each type of distributional requirement. Numbers in parentheses in the second column are standard errors of the estimated values of β. The third column lists the number of selection sets used within each category, while the fourth and fifth columns represent the probability and odds, respectively, that a student selected a course with an A– average versus a course with a B average.

Distribution Requirement	β (se)	Sample Size	Probability Selected	Odds
Civilizations	1.36 (0.98)	27	0.72	2.59
Social Sciences	1.23 (0.37)	139	0.70	2.36
Natural Sciences	−0.19 (0.40)	90	0.47	0.87
w/o chem and bio	1.34 (0.49)	57	0.72	2.55
Arts and Literature	1.28 (0.92)	30	0.71	2.45
Quantitative Reasoning	0.78 (0.46)	54	0.63	1.72
Languages	6.23 (5.11)	4	0.99	78.3
All distribution requirements	1.15 (0.23)	311	0.69	2.24
Selection by Instructor	0.94 (0.36)	197	0.66	1.93
Combined	1.09 (0.20)	508	0.68	2.14

One explanation for this discrepancy might be that natural science courses elicit the best from students, and considerations of higher grades therefore play little or no role in the selection decisions of students who take these courses. Another, and perhaps more realistic, explanation for this negative coefficient is that 33 of the 90 selections identified for natural science courses contained courses required in the premed curriculum (e.g., introductory biology and chemistry, and organic chemistry). Notoriously low mean grades are assigned in several of these courses, and it is possible that students examined the mean grades of these courses simply out of curiosity before taking them. Accepting the latter explanation, and remembering that interest in this analysis focuses on the effects of grading policies on students who have a choice in the courses they take, the selection sets used to estimate the regression coefficient for the natural science courses were redefined so as to exclude premed courses. In other words, all selection sets containing any of the premed courses (even if not the course taken by the student) were pared from the analysis. Based on the remaining 57 selections, the maximum likelihood estimate of β for natural science courses was 1.34, which is similar to the values obtained for the other categories of educational requirements.

The fourth column of Table 1 lists the estimated probability that a student choosing between a course with an A− average and a B average would choose the course that had an A− average. The fifth column provides the corresponding odds.

The last two rows of the table summarize results for the effect of expected mean grade on student course selection decisions for courses selected by distribution requirement and for courses selected by instructor. Values appearing in the third-to-last row were estimated using the concatenation of selection sets obtained for the six academic areas (excluding premed courses). Note that the estimated values of β are relatively stable across each selection category.

Of course, these results are based on examining only a small proportion of the student course selection decisions that occurred during the course of the DUET experiment. It is therefore important to examine the extent to which these course selection decisions were similar to those not captured in the above analysis. We examine this question next.

EFFECTS OF SAMPLE SELECTION

The analyses in the preceding section were based on a relatively small proportion of the student course selection decisions made during the course of the DUET experiment. Of the roughly 5,700 undergraduates enrolled at Duke during the 1998–1999 academic year, slightly over 2,000 undergraduates opted to participate in the DUET survey. These 2,000 students made approximately 16,000 course selection decisions during this period, and of these, only about 500, or 3%, were captured in the sampling process used in these analyses.

To examine whether the course selection decisions captured in these analyses are typical of the larger population of course selection decisions made by students at Duke, student opinions about courses included in the analyses were compared to student opinions collected for all other courses surveyed in the DUET experiment. Indeed, it was for this reason that items querying students' awareness of grading policies before enrolling in a course (item 23) and students' opinions on the extent to which grading policies affected their enrollment decisions (item 24) were included on the survey. Unfortunately, the DUET experiment was terminated by the Duke University Arts and Sciences Council[4] after only two semesters. Because course evaluation data for courses selected in these analyses were scheduled to be collected for the first time in the third semester of the experiment, the termination of the experiment after two semesters meant that no course evaluation data were collected for any of the courses captured in the preceding section.

To overcome this problem, a follow-on survey was conducted after the conclusion of the formal DUET experiment. In the follow-on survey, an email version of the DUET questionnaire was sent to 250 students who had taken a course for which they had previously examined the mean grade. Survey responses were requested only for the course in question, and students were informed that the purpose of the follow-on survey was to assess the stability of their responses over time. Sixty-five students responded to the email version of the survey.[5]

There are two ways that responses from the follow-on survey can be used to compare courses captured in a selection-set analysis to courses that were not. First, the distribution of responses to DUET items can be compared across the two groups. Significant differences in these distributions signify that courses included

---

[4]The original experiment had been approved for a three-year period by a subcommittee of the Arts and Sciences Council, the Academic Affairs Committee.

[5]Of course, none of the 250 students had actually completed the original DUET survey for the course that was the subject of the email follow-on, because the spring DUET survey was completed for courses taken the previous fall. However, none of the students included in the follow-on survey commented on the fact that they had not previously completed the survey for the course in question. Five students did not provide item responses but instead provided general comments. Two students did not complete later items on the survey.

in the selection sets were somehow viewed differently by students. If present, such differences would argue against the extension of conclusions from this analysis to the broader population of course selection decisions made by Duke students. Second, responses to DUET items on the follow-on survey can be used to adjust the estimates of the regression coefficients summarized in Table 1 to account for variations in the responses of the two groups.

Intuitively, an item that might play an important role in determining the influence of grading policy on a student's decision to enroll in a course is the student's use of that course in satisfying graduation requirements. At Duke, students take courses to satisfy major requirements, distributional requirements, or as electives that count toward their 34-course minimum requirement. A student's purpose for taking a course was ascertained in item 2 of the DUET survey.

Figure 2 depicts the observed proportions of students taking courses for each of these purposes. The top panel reflects the proportions observed for all 11,521 responses obtained for this item during the DUET experiment. The lower panel depicts the same proportions obtained from the respondents in the follow-on email survey. These two distributions appear to be nearly identical, and in fact, the χ^2 statistic for testing whether the distribution in the lower panel differed from the distribution of responses observed in the upper panel is highly *insignificant* ($p = .54$). Thus, there is no evidence of any systematic difference in the reasons students took courses between those students and courses included in the analyses of the previous section and those students and courses whose course selection decisions were not included in those analyses.

Another item that might reasonably be assumed to affect the magnitude of the influence of grading policies on students' decisions to select courses is prior student awareness of how a course is likely to be graded. Prior awareness of how stringently a course would be graded was probed in item 23 of the DUET survey, which asked, "On a scale of 1–5, with 1 being 'completely unaware' and 5 being 'completely aware,' how aware were you of how this course would be graded when you enrolled in it?" The distributions of responses to this question for all courses surveyed during the DUET experiment

FIGURE 2

Comparison of students' purposes for taking courses included in the sampling procedure to all courses surveyed in the DUET experiment. The top panel reflects responses received for item 2 from all students participating in the DUET experiment, and the lower panel, those responses obtained from students during the follow-on email survey. Only students whose courses were included in the analyses of the previous section were included in the follow-on survey. Courses taken to satisfy a major requirement are reflected in the leftmost column; courses taken to satisfy a distributional requirement appear in the center column; and courses taken as electives are depicted in the rightmost column.

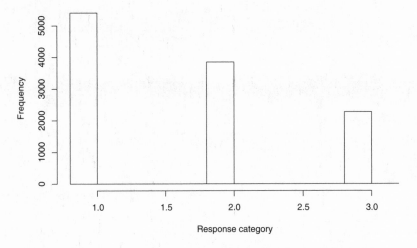

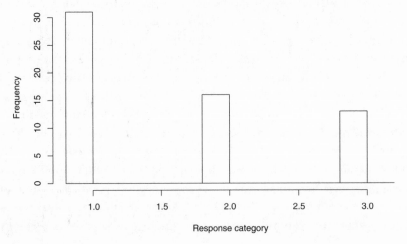

and for the students who responded to the email follow-on survey are depicted in Figure 3.

The most notable difference between the top and bottom panels of Figure 3 is the absence of "Not Applicable" responses in the lower plot. While 4% of the total DUET responses to this item were "Not Applicable," none of the respondents to the follow-on survey provided this response. However, only two or three "Not Applicable" responses were expected in a survey of this size anyway, and so this difference is not particularly severe. The χ^2 statistic for testing differences between these distributions is again not significant ($p = .58$). As with item 2, there is little evidence of any systematic difference between students' prior knowledge of course grading policies in courses included in the selection-set analyses and a randomly selected course observed in the DUET experiment.

While items 2 and 23 provide tangential insight into possible differences between the influence of grading policies on students' enrollment decisions for courses captured in the analyses of the previous section versus those that were not, the most definitive evidence for such selection biases is expected from student responses to item 24, which asked, "How much did your knowledge of how the course would be graded positively affect your decision to enroll in it?" Assuming that students in the two populations were equally forthright in answering this question, responses to this item provide a direct measure of the generalizability of conclusions from the selection-set analyses to the broader population of course selection decisions made at Duke.

Histogram estimates of the distributions of student responses to this item for the two groups are depicted in Figure 4. These plots display more disparity than previously noted for either item 2 or 23, with the most notable difference again occurring in the "Not Applicable" category. However, the χ^2 test for differences between the populations is again not significant ($p = .21$).

Despite the similarity in responses collected for items 2, 23, and 24, it is still worth examining the possible impact of differences in these response patterns on the conclusions reached in the previous section. To further investigate systematic variations in the effects of grading policies on student course selections as

FIGURE 3

Comparison of students' prior knowledge of course grading policies. The first five columns depict the proportion of responses obtained in each numbered category, while the sixth column depicts the number of students who offered "Not Applicable" response. The top panel represents the distribution of responses obtained for all courses surveyed during the DUET experiment, the lower panel the follow-on email survey.

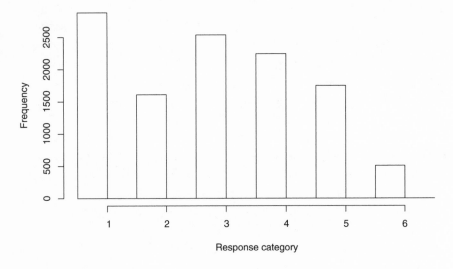

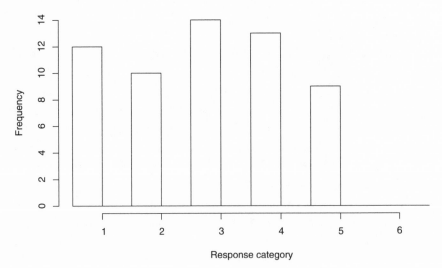

FIGURE 4

Comparison of the effect of grading policy on students' enrollment decisions. Responses collected during DUET experiment are displayed in the top panel, responses collected during follow-up email survey appear in the lower panel. "Not Applicable" responses are shown in column six.

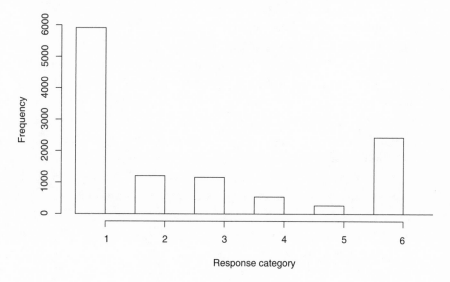

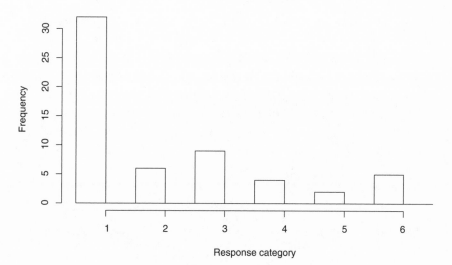

a function of response to item 24, the responses of the students who completed the follow-up email survey were used to estimate values of β in (6.1) for each response category in item 24.[6] The posterior means (and standard deviation) of β estimated for each response category from this procedure were 1.36 (0.39) for the first response category ("No effect or negative effect"), 0.79 (0.83) for the second response category ("Slight effect"), 0.91 (0.61) for the third response category ("Moderate effect"), 0.61 (.89) for the fourth response category ("Significant effect"), and 1.09 (.89) for the fifth response category ("Very significant effect"). The posterior mean of β for "Not Applicable" responses was 1.32 with a standard deviation of 0.90. Note that the largest, most positive values of β were estimated for the "Negative effect or no effect" and "Not Applicable" categories. The posterior mean of β for the total population, obtained by averaging the values of β obtained for each response category according to the population proportions observed for that response category in the DUET experiment, was 1.20 with a standard deviation of 0.27.

These results are interesting from several perspectives. First, comparatively few students (40%) stated that grading policy had any impact on their enrollment decisions. This contrasts sharply with student opinions expressed during the achievement index controversy and with most academic folklore. In addition, estimates of β tended to be most positive for students who stated that grading policy had either no effect, a negative effect, or did not apply to their enrollment decisions. This suggests a certain degree of deception on the part of students, who apparently did not wish to admit that grading policies played some role in their enrollment decisions. This possibility finds further support from the substantial number of students (21%) who responded to item 24 in the "Not Applicable"

---

[6] A latent variable model was used in this estimation procedure to account for missing responses to item 24 for courses included in the selection sets. For each course taken and included in a selection set, and for which the response to item 24 was missing, a latent variable was introduced for the unknown survey response. The marginal distribution of these latent variables was assumed to have a multinomial distribution with probabilities drawn from the posterior distribution on the response categories obtained using the observed responses from the email respondents in conjunction with Jeffrey's (noninformative) prior. The posterior distribution on each value of β was obtained by integrating the posterior distribution over the values of these latent variables.

category. After all, a "Not Applicable" response seems inappropriate for this item, since the first response category of "Negative or no effect" covers the event that grading policy had no effect on their enrollment decision. In any case, the adjusted value of β, adjusted for student response to item 24, is *larger* than the unadjusted value based only on courses included in the analyses of the previous section.

Nonrandom sampling of courses included in the data set undoubtedly led to some bias in the estimates of β in (6.1). However, the direction of this bias is not clear, and results based on the follow-on survey suggest that the influence of this sampling bias was negative; that is, the influence of expected grade on course selection decisions is likely larger than estimated in these analyses. Support for the hypothesis of a negative bias is also garnered from the following demographic features of the DUET participants.

One of the largest differences in the demographic attributes of students included in the selection-set analyses versus those who were not was academic year. A disproportionate number of first- and second-year students participated in the DUET survey, and a disproportionate number of these were represented in courses included in the selection sets. Underclassmen tended to report fewer "Not Applicable" responses than did third- and fourth-year students, and because "Not Applicable" responses were associated with larger estimates of β, the disproportionate number of first- and second-year students used to estimate β suggests a negative bias. Furthermore, first- and second-year students typically have less knowledge of expected course grading policies than do more senior students, and so have less opportunity to allow this knowledge to affect their course enrollment decisions.

To summarize the effects of potential sample-selection biases in these analyses, several points should be emphasized. First, as in most surveys, a thorough examination of the biases attributable to nonrandom sampling is not possible because the data required for this task was, by definition, not collected. This being said, the data available for examining differences between the sampled population and the nonsampled population suggest that these groups differed only slightly in the attributes that would most likely signal a response bias. Approximately the same proportions of courses were taken to satisfy major, distributional, and elective requirements in both groups; prior knowledge of course grading policies was similar in both groups; and student-reported effects of grading policies on course selection decisions were similar in both groups. Where there were noticeable differences in these attributes, adjusted estimates of the magnitude of the effect of grading policy on student course selection decisions were larger than the unadjusted estimates. Thus, although the precise impact of selection biases on the results of the last section cannot be determined, it seems plausible that the estimates obtained there provide at least a first-order approximation to the true magnitude of these effects.

EXTENSION TO SELECTION DECISIONS BETWEEN ACADEMIC FIELDS

In previous sections, the effects of grading policies on student course selection decisions were examined for students selecting among instructors of the same course and among courses that satisfied the same educational requirements. Extending the results from these analyses to selection decisions that students make when selecting courses from different academic fields is equally important, but is difficult to accomplish using data from the DUET experiment due to the presence of several confounding factors that make a direct examination of such selections inappropriate. Effects of grading policies on course selection decisions made for selection of courses in different academic fields must therefore be based on the extrapolation of the results obtained for selection of instructors for the same course and selections of courses satisfying the same academic requirements.

The validity of this extrapolation depends on the stability of the estimates of the effects of grading policies on student course selections across the various purposes for which courses were taken. That is, if estimates of β in (6.1) do not vary substantially across the categories of course selection decisions that a student makes, then the extrapolation of the magnitude of these effects to course selection decisions made among courses taken in different academic disciplines seems reasonable. If estimates of β fluctuate wildly among the disciplines from which students have selected a course, then it is not.

Evidence presented in the first section of this chapter suggests that the influence of expected course grades on students' decisions to enroll in a course does not vary substantially between fields. Table 1 shows that estimates of β obtained for course selection decisions made within the six categories of distributional requirements (excluding premed courses) range from 0.78 to 6.23. The estimate of 6.23 was based on only four observations for the Foreign Languages category, and was estimated with a standard error of 5.1. Excluding this estimate, the range is narrowed to 0.78 to 1.36, with standard errors ranging from 0.46 to 0.98. Furthermore, the smallest estimate, 0.78, corresponds to course selection decisions made within the Quantitative Reasoning category, and this category may have been subject to some of the same biases that affected Natural Sciences when premed chemistry and biology courses were included. In the case of Quantitative Reasoning courses, premed students, as well as students from several academic majors, are required to take one or two semesters of introductory calculus, and the Department of Mathematics at Duke University imposes constraints on the mean grades awarded in its calculus sections. As a result, the estimate of β obtained for the Quantitative Reasoning category may be biased downwards because of the inclusion of these required courses. Despite this, the value of β and associated standard error estimated for Quantitative Reasoning courses are still relatively consistent with the values estimated for the other distributional requirements, which range from 1.23 to 1.36.

The estimated value of β for students selecting from among instructors of the same course was 0.94 with a standard error of 0.36. This, too, is consistent with the range of values estimated for distributional requirements.

The final issue that must be considered before extrapolating to students' selection of courses taken in different academic fields are the differential effects attributable to students' reasons for taking a course. To assess the interaction between students' reasons for taking a course and the influence of expected grading policy in their enrollment decisions, the data from the follow-on email survey were used in a latent variable analysis similar to that described above for DUET item 24 to obtain estimates of β for each category of response in item 2. The resulting estimates of β for courses taken to satisfy major requirements, distributional requirements, and electives were 1.16 (0.26), 1.07 (0.32), and 1.15 (0.32), respectively. Note that these estimates, too, are relatively stable within categories and are similar to the other estimates displayed in Table 1.

All of these considerations suggest that the effects of grading practices on student course selection decisions are relatively unaffected by field of study and reason for taking a course.

Now suppose that from a total of 34 courses taken by an average Duke student, ten courses are taken to satisfy major requirements and twelve courses are taken to satisfy distributional requirements. This leaves a dozen courses that may be chosen as electives. For each of these electives, a student may choose to take a course in the humanities, social sciences, or natural sciences and mathematics. Let p_H denote the hypothetical probability that a student would choose a humanities course if the grading practices in all academic divisions were the same, let p_{SS} denote the corresponding probability that a student would choose a social science course if the grading policies in all academic divisions were the same, and let p_{NSM} denote the probability that a student would choose a natural science or mathematics course if the grading policies in all academic divisions were the same. Note that in reality, p_H, p_{SS}, and p_{NSM} are not directly observable because the grading policies in different academic divisions are not the same. The goal in what follows is to estimate these probabilities and to compare them to the proportions of classes that students do take in each division under current grading practices.

Mean course grades and the proportions of courses taken in each academic division by students who participated in the DUET survey are listed in Table 2.[7] To relate the proportions in this table to the

---

[7] Engineering courses have been excluded from this analysis for reasons cited in Chapter 2.

TABLE 2

Proportion of courses taken and mean course grades listed according to students' purpose for taking a course and academic division.

Purpose of Course	Humanities	Social Sciences	Natural Science and Math
Major Requirements:			
Proportion:	0.43	0.33	0.24
Mean grade:	3.53	3.35	3.08
Distribution Requirements:			
Proportion:	0.51	0.30	0.19
Mean Grade:	3.54	3.39	3.11
Electives:			
Proportion:	0.50	0.27	0.23
Mean Grade:	3.54	3.40	3.05

probabilities p_H, p_{SS}, and p_{NSM}, note that under the assumptions discussed above, the proportion of classes taken in each division may be expressed as a product of p_H, p_{SS}, and p_{NSM} and a factor proportional to (6.1), with, say, $\beta = 1.15$. That is, the proportion of elective humanities courses taken, 0.50, is assumed to be proportional to

$$p_H \times \exp(1.15 \times 3.54);$$

the proportion of social sciences courses taken, 0.27, is assumed to be proportional to

$$p_{SS} \times \exp(1.15 \times 3.40);$$

and the proportion of natural science and mathematics courses taken, 0.22, is assumed to be proportional to

$$p_{NSM} \times \exp(1.15 \times 3.05).$$

Solving these equations under the constraint that the probabilities sum to one leads to $p_H = 0.41$, $p_{SS} = 0.26$, and $p_{NSM} = 0.33$.

How significant are the differences between the estimated values of p_H, p_{SS}, and p_{NSM} and the observed proportions of students taking elective courses in each academic division? Averaged over the entire undergraduate population at Duke, these differences suggest that students would occupy 7,000 more seats in science and mathematics courses, and approximately the same number fewer seats in humanities courses, if grading policies were more equitable. In other words, if differences in grading policies between divisions were eliminated, the average undergraduate at Duke would probably take, on average, 4.0 natural science and mathematics electives instead of the 2.8 electives that they currently do. This would represent a nearly 50% increase in the number of natural science courses taken.

A 50% increase in elective natural sciences courses would have substantial repercussions on the allocation of the university's faculty teaching resources. Assuming an average class size of 25 students, a 1.2 course increase per student would require the addition of approximately 75 science courses per year at Duke (and other comparably sized universities). In terms of faculty positions, this equates to the addition of about 20 to 25 full-time science faculty in an institution that has a total faculty size of about 500, and, of course, the potential loss of approximately the same number of humanities positions. And these changes correspond only to shifts that would occur as a result of changes to *elective course* enrollment patterns.

Shifts in student enrollment patterns would undoubtedly also occur as a consequence of students changing majors and choosing to satisfy different distributional requirements. For example, many premed students at Duke now opt to major in psychology because it is easier to maintain a high GPA in psychology than it is in chemistry or biology. If grading policies were more equitable, it is likely that many of these students would instead choose natural science majors. Each student that converted to a science major would take approximately ten additional science courses to satisfy major requirements, and would likely take ten or so fewer social science courses. Similar conclusions apply for distributional requirements; if grading policies were more uniform across disciplines, the analysis presented in the previous section suggests that more students would choose to satisfy Natural Science or Quantitative Reasoning distributional requirements than currently do. Although the DUET data do not provide an appropriate mechanism for assessing the number of students who would change majors or satisfy different distributional requirements, the analysis of the effects on elective courses, together with the data provided in Table 2, suggests that differences in grading policies probably play an equally significant role in these decisions.

Conclusions

The influence of grading policies on student course selection decisions is substantial; when choosing between two courses within the same academic field, students are about twice as likely to select a course with an A− mean course grade as they are to select a course with a B mean course grade, or to select a course with a B+ mean course grade over a course with a B− mean course grade. This fact forces instructors who wish to attract students to grade competitively, but not in the traditional sense of competitive grading.

For instructors, the implications of this finding are quite alarming, particularly when considered in conjunction with the findings of previous chapters. An instructor who grades stringently is not only less likely to receive favorable course evaluations, but is also less likely to attract students (another indication of poor teaching?). Because most departments are hesitant to devote teaching resources to undersubscribed classes, this means that stringently grading instructors are also less likely to have the opportunity to teach specialized courses in their academic area. And of course, to the extent that personnel decisions at an institution are based on teaching effectiveness, or at least on the institution's perception of teaching effectiveness, stringently grading faculty are less likely to be promoted, to receive salary increases, or to be tenured.

At the institutional level, differences in grading policies among academic divisions result in substantial decreases in natural science and mathematics enrollments, and artificially high enrollments in

humanities courses. This shift in enrollment causes a disproportionate allocation of resources to humanities departments at the expense
of science departments.

Nationally, the effects of inequitable grading practices on student course selection decisions are extremely difficult to quantify.
How does one measure the costs of scientific illiteracy? How has
public discourse on issues ranging from stem cell research to genetic
alteration of food products to discussion of missile defense technology to environmental protection policies been affected? To what
extent has the ability of the average college graduate to function
in an increasingly technological society been compromised? The
answers to these questions cannot be addressed using data collected
during the DUET experiment. What can be said, however, is that
the general level of scientific competence in America has been
diminished simply because universities have not adopted more
consistent grading policies.

7   Grading Equity

This chapter documents disparities in grading practices among disciplines by examining average differences in the grades received by the same students in different fields. Natural science, mathematics, and economics courses are found to be the most stringently graded disciplines; humanities courses are the most leniently graded.

Effects of these grading disparities on student assessment are severe. For many Duke students, the grading policies used by their instructors were nearly as important in determining their GPA and class rank as was their academic performance.

Two methods of adjusting grades to account for these differences are illustrated.

FOR STUDENTS, THE CURRENCY OF ACADEMIA IS THE grade. As the only tangible benefit that students receive for performing well in their courses, grades provide the primary mechanism available to the faculty for maintaining academic standards. In a very real sense, professors pay grades to students in return for mastery of course material, and students barter these grades for jobs or entrance into professional or graduate school.

In the early 1960s, student grades, like the dollar, were taken off the gold standard. Over the course of the next thirty-five years, inflation led to significant decreases in the value of both college grades and the dollar, and by the late 1990s, a grade of B generally represented the same "average" performance that a C had in 1960.

Some might argue that this inflation has caused no damage, and that like our economy, postsecondary education expanded and prospered during this period. The fact that an average student now receives a B instead of a C has not intrinsically devalued our educational system. A grade of C represents nothing more than an ordered categorical response; the definition of the categories is completely subjective.

In reality, however, grade inflation did damage our educational system, but only indirectly. It damaged the system by drawing attention away from a much more serious flaw in the academic monetary system: It masked increasingly wild fluctuations in the exchange rates between academic disciplines.

Causal links between the resulting grading disparities, student enrollment patterns, and distortions in faculty teaching evaluations have already been established. In this chapter, evidence of fluctuations in the exchange rates of grades between disciplines is presented. Much of this evidence derives from an examination of pairwise differences of student grades received by the same students in courses taught in different fields. By examining such pairwise differences, effects attributable to student characteristics cancel, making results from this type of analysis particularly compelling.

After demonstrating that grading practices do differ between disciplines, two methods for adjusting student GPAs to compensate for these differences are presented. The products of these methods—adjusted GPAs and adjusted course grades—may be used in a variety of ways.

Adjusted grades can potentially be used to replace or supplement standard grade summaries on college transcripts, either as an option presented to students individually or universally for all students. Alternatively, adjusted grades or grade ranges can be used to provide feedback to instructors regarding their own grading policies. Such feedback mechanisms are sorely lacking at most universities and might, by themselves, result in substantial reductions in the disparities of grading policies across disciplines and instructors. Finally, by providing an improved estimate of the average student achievement level within classes, these methods provide a yardstick for modifying university-imposed grading standards according to the ability levels of students within courses. Further discussion of this and other reform strategies is presented in Chapter 8.

Before examining variation in grading practices among academic fields, it is important to note that disparities among the grading practices of individual instructors within academic fields often dwarf the disparities observed between fields. However, because instructor differences in grading practices are, by their nature, relatively specific, little attention is paid here to the examination of such differences. In a sense, this is unfair to stringently grading professors from leniently grading disciplines, or, depending on your perspective, leniently grading professors from stringently grading disciplines.

DIFFERENTIAL GRADING STANDARDS

H istorically, Goldman and colleagues were among the first researchers to systematically investigate differences in grading standards across academic disciplines. Their initial foray into this arena involved a study that focused on predicting the effects that various college majors had on student GPAs [GSHF74]. Although the approach used in this study is only tangentially related to more methodologically sound investigations that were to follow, it is worth reviewing for its insights into the mechanisms that cause differences in grading practices among disciplines.

The basis for the analyses in [GSHF74] were regression models in which GPAs of students enrolled at the University of California, Riverside, were predicted using students' high school GPAs, math and verbal SAT scores, and college major. Separate regression models were estimated for students in each of twelve major fields. The fields studied were Anthropology, Biochemistry, Biology, Chemistry, Economics, English, History, Mathematics, Political Science, Psychology, Sociology, and Spanish.

Interpreting the results from these regression functions was complicated by a technical difficulty involving the fact that the fitted regression surfaces were not parallel. To overcome this complication, Goldman and colleagues divided students into three ability levels according to GPA. Students with GPAs less than 2.7 (B−) were assigned to Group I, students with GPAs between 2.7 and 3.3 (B+) were assigned to Group II, and students with GPAs greater than 3.3 were placed into Group III. For students in each academic major, mean high school GPA, mean math SAT score, and mean verbal SAT score were calculated for each of these three GPA groups. These values were then used as explanatory variables in regression equations in which college GPA was the dependent variable.

The GPAs predicted for each group of students from these regression equations are summarized in Tables 1 and 2. Table 1 lists the predicted GPAs of students from each major averaged over all fields in which they might have majored. Table 2 provides the complementary data; the entries in this table represent the mean GPA that could be expected had every student at UC Riverside majored in the given field. Thus, the first entry in Table 1, 2.61, is the GPA predicted for a randomly selected anthropology major from Group I when this student is graded using the average grading standard applied at UC Riverside. Likewise, the first entry in

TABLE 1

Predicted GPAs of students from 12 academic majors. The figures in this table represent the GPA predicted for students in each department had they instead majored in a department that employed an "average" grading scheme. Extracted from Tables 3 through 5 of Goldman, Schmidt, Hewitt, and Fisher [GSHF74].

Field	GPA ≤ 2.7	2.7 < GPA ≤ 3.3	3.3 < GPA
Anthropology	2.61	2.79	3.05
Biochemistry	2.70	2.96	3.14
Biology	2.71	2.89	3.06
Chemistry	2.91	2.95	3.18
Economics	2.66	2.82	2.91
English	2.76	2.91	3.05
History	2.67	2.79	2.99
Mathematics	2.80	2.96	3.18
Political Science	2.46	2.77	3.11
Psychology	2.68	2.89	2.96
Sociology	2.47	2.75	2.89
Spanish	2.38	2.67	2.99

Table 2, 2.55, represents the predicted mean GPA of all students in Group I had all these students majored in anthropology.

Systematic differences in both the predicted GPAs of students from different majors and the GPAs predicted for equally able students as they change major are apparent in these tables. Goldman, Schmidt, Hewitt, and Fisher argue that these two trends provide evidence that Helson's theory of adaption level [Hel47, Hel48] operates when instructors set their grading policies. According to this theory, instructors anchor their judgment of a student's performance by comparing that student's performance to the performance of other students in their classes. From Tables 1 and 2, we see that the mean grades awarded in disciplines with less-able students are actually higher than the mean grades awarded in disciplines with better students. The theory of adaptation level provides a partial explanation for this phenomenon.

TABLE 2

Predicted GPAs for students in each GPA group when majoring in one of the twelve academic majors listed. Extracted from Tables 3 through 5 of Goldman, Schmidt, Hewitt, and Fisher [GSHF74].

Field	GPA ≤ 2.7	2.7 < GPA ≤ 3.3	3.3 < GPA
Anthropology	2.55	2.74	2.93
Biochemistry	2.49	2.73	2.96
Biology	2.51	2.75	2.99
Chemistry	2.49	2.75	3.02
Economics	2.51	2.58	2.66
English	2.72	2.91	3.11
History	2.76	2.94	3.13
Mathematics	2.45	2.75	3.05
Political Science	2.81	3.02	3.25
Psychology	2.78	2.91	3.06
Sociology	2.78	2.90	3.04
Spanish	3.05	3.19	3.33

As an aside, the observation that grade distributions are highest in those disciplines that have the least-able students has important implications when considering policy changes to correct for disparities in grading practices. If ability levels of students do vary substantially among disciplines, then applying a uniform grading constraint would have the unintended effect of forcing instructors to apply *different* standards when assigning student grades.

While the Goldman, Schmidt, Hewitt, and Fisher study provided one of the first glimpses into the disparities of grading policies among academic disciplines, it also suffered from two methodological shortcomings. First, the results of the study relied on the prediction of college GPA using regression functions with only three covariates: high school GPA (HSGPA), SAT math, and SAT verbal score. As the authors themselves point out,

... it is very likely that GPA is determined by many factors other than the abilities measured by HSGPA and SAT. Perhaps, the fields with "easier" grading standards may have students who are more diligent, creative, or possess more of other unmeasured traits which are positively related to GPA. In such a circumstance, it would be unfair to characterize a field as having "easy" grading. Unfortunately, we have no information on these personality characteristics and must base our findings upon HSGPA and SAT measurements, as other investigators have done [GSHF74, 355].

The second problem with their study was that student GPAs were calculated using all courses taken. Because major courses constituted only about one-third of the courses taken by students in the study, there was severe attenuation of the estimates of the effects attributable to grading differences between majors. As the authors state, their "results may actually be quite *conservative*" [GSHF74, 355].

Many of the methodological problems encountered in the Goldman, Schmidt, Hewitt, and Fisher study were resolved in a later study by Goldman and Widawski [GW76]. In this later study, pairs of grades obtained by the same student in different departments were used as the basis for between-department grade comparisons. By examining pairwise differences in grades from the same student, individual student characteristics, like diligence and creativity, were largely eliminated as possible confounding factors in subsequent analyses. So, too, was the reliance on a simplistic regression model for predicting GPAs across academic disciplines.

In the Goldman and Widawski study, one-quarter of all first- and third-year students at the University of California, Riverside, were randomly sampled. All courses taken by these students were placed into one of the following 17 academic categories: Anthropology, Art, Biology, Chemistry, Economics, English, Ethnic Studies, Foreign Language, Geology, History, Mathematics, Philosophy, Physics, Political Science, Psychology, Sociology, and Urban Studies. From these 17 categories, the authors estimated the 136 possible pairwise comparisons of grading practices between fields by examining the difference of grades received by students who had taken a course in both fields. For example, the difference in grading policies between Anthropology and Art was estimated by identifying all students who had taken a course in each field, and for each such student, subtracting the grade received in Anthropology from the grade received in Art.

Goldman and Widawski provide a table containing the mean pairwise differences for each of the 136 comparisons estimated in this way. Of these differences, 52 were reported to be statistically significant at the 5% level. In addition, the pairwise differences were judged to be nearly additive, meaning that the addition of the pairwise difference between Anthropology and Art, and Art and Biology nearly equals the pairwise difference between Anthropology and Biology (in this case, $-.34 + .72 = .38 \approx .28$). Only 11% of the variation in the pairwise differences was accounted for by nonadditivity. This probably means that study conclusions were not seriously affected by selection biases that might have caused special groups of students to contribute disproportionately to particular comparisons.

By averaging the pairwise differences in mean grades across academic fields, Goldman and Widawski obtained a "grading index" for each discipline. For instance, the grading index for Anthropology was obtained by averaging the pairwise differences of grades obtained in Anthropology with the grades assigned to the same students in all other fields. The grading indices so obtained are displayed in Table 3. This table shows that, for example, students who took biology courses received, on average, a one-half lower letter grade than students who took courses in political science, while students enrolled in ethnic studies or urban studies received a one-third higher letter grade, on average, than students who opted instead to take a course in psychology.

To test whether the theory of adaptation level applied also to these data, Goldman and Widawski computed the correlation between the grading indices and high school GPAs, math SAT scores, and verbal SAT scores for students who had majored in each field. These correlations were -0.67, -0.64, and -0.65,[1]

---

[1] A typographical error in Table 2 of Goldman and Widawski led to the omission of the negative sign in front of this correlation.

TABLE 3

Mean pairwise grading differences by academic fields. These grading indices
were obtained by averaging pairwise differences of grades obtained by the
same student in two academic fields. The second column summarizes results
from Goldman and Widawski [GW76] using student grades awarded at the
University of California, Riverside. Column 3 contains the comparable results
from Strenta and Elliot [SE87] using grades from introductory courses at Dartmouth College in 1983, while Column 4 contains the corresponding results for
all courses. Column 5 summarizes results of Elliot and Strenta [ES88] using
grades awarded at Dartmouth in 1986. In the fifth column, the row labeled
"Art" was originally labeled "Art/Visual Studies," "Chemistry" was listed as
"Chemistry/Biochemistry," "Geology" was "Earth Science," "French" was
"French & Italian," and "Mathematics" was "Math/Computer Science."

Field	G&W Index	S&E (Intro) Index	S&E (All) Index	E&S Index
Anthropology	−0.09	−0.05	−0.02	0.00
Art	+0.22	−0.10	+0.09	+0.06
Asian Studies	—	—	—	+0.12
Biology	−0.53	−0.27	−0.28	−0.32
Chemistry	−0.36	−0.40	−0.32	−0.35
Comparative Lit.	—	—	—	+0.31
Drama	—	—	—	+0.37
Economics	+0.18	−0.35	−0.35	−0.44
English	−0.10	+0.07	−0.01	+0.05
Engineering	—	—	—	−0.16
Ethnic Studies	+0.38	+0.42	+0.28	—
Foreign Language	+0.06	—	—	—
French	—	+0.38	+0.23	+0.08
Geology	−0.30	−0.16	+0.18	+0.02
Geography	—	—	—	−0.16
German	—	—	—	−0.07
Government	—	+0.04	−0.04	−0.19
Greek & Latin	—	—	—	+0.05
History	+0.05	+0.23	+0.01	−0.07
Mathematics	−0.07	−0.25	−0.24	−0.37
Philosophy	+0.17	+0.01	+0.02	−0.07

(continued on next page)

TABLE 3 *(continued)*

Field	G&W Index	S&E (Intro) Index	S&E (All) Index	E&S Index
Physics	−0.23	−0.25	−0.16	−0.25
Policy Studies	—	—	—	+0.09
Political Science	−0.03	—	—	—
Psychology	+0.03	+0.01	−0.08	−0.16
Religion	—	+0.17	+0.11	+0.03
Russian	—	—	—	+0.02
Sociology	+0.29	+0.10	+0.31	+0.25
Spanish	—	+0.38	+0.25	+0.20
Urban Studies	+0.38	—	—	—

respectively. Negative correlations between the grading indices and external measures of student ability indicate that higher grading indices were associated with fields whose students had lower high school GPAs, lower math SAT scores and lower verbal SAT scores. This fact provides further support for the hypothesis that instructors in fields that attract students of comparatively lower ability tend to grade leniently, while instructors in fields that attract higher-ability students tend to grade more stringently.

A decade after the publication of Goldman and Widawski's study, Strenta and Elliot [SE87] and Elliot and Strenta [ES88] replicated this study design using student data collected at Dartmouth College. They also extended the Goldman and Widawski design in several ways. In [SE87], they computed pairwise differences using all pairs of courses taken by the same student, as had been done by Goldman and Widawski, and also by using only pairs of introductory courses. Their motivation for computing grading indices using only introductory courses was to eliminate possible effects of course level. In [ES88], they extended the analysis still further by examining more departments and estimating pairwise differences between grades awarded to the same student in courses taught within the same department. Complete tables of pairwise differences and their corresponding significance levels can be found in the original articles, and, like the previous results of Goldman and Widawski, the numbers in these tables are nearly additive. Summary grading indices based on the Elliot and Strenta studies are provided alongside Goldman and Widawski's estimates in Table 3.

There is a striking similarity between the mean pairwise differences reported by Goldman and Widawski in 1976 using student data collected from a moderate-sized public university and those reported by Strenta and Elliot more than a decade later using student records collected at a small, moderately selective private college. The correlations between the second column of this table, generated using the University of California, Riverside, data, and the remaining columns of the table, based on data from Dartmouth College, range from 0.43 to 0.63. Furthermore, there is a consistent trend for humanities departments to grade leniently (positive grade indices), for social sciences departments to grade in an approximately neutral manner, and for natural sciences and mathematics departments to grade stringently. The notable exceptions to this trend are economics courses, where grading tends to be quite stringent, and sociology courses, where grading tends to be lenient.

Strenta and Elliot also examined correlations between departmental grading indices and external summaries of student abilities. For their 1983 data, the correlation between the grading indices (using all courses) and the average student SAT score (verbal + math) was −0.76. The corresponding result using the Goldman and Widawski indices was −0.68. For their 1986 data, the correlations between the grading indices and average SAT score was −0.60; with high school class rank it was −0.61. These correlations provide further support for the operation of Helson's theory of adaptation level in determining grading policies. Departments that attracted the most capable students graded stringently, while departments that attracted less-capable students graded more leniently.

Of course, alternative explanations can easily be constructed to account for the negative correlations between departmental grading indices and external measures of student ability. The most plausible is that discipline-specific abilities underlie observed differences in grading stringency between departments. Strenta and Elliot, like Goldman and Widawski before them, recognized this possibility and in their 1987 article summarized this concern by stating that "it is readily conceivable that students who get good grades in performing arts courses, for example, and poorer grades elsewhere might have special talents for their art that are relatively independent of general ability—that is, the higher grade reflects better performance and not merely less stringent grading" [SE87, 289].

This distinction is important. If differences in grading standards can be attributed to "special talents" possessed by some students and not others, then policy changes enacted to correct for these differences would clearly be both unfair and inappropriate.

To investigate the extent to which grading differences can be attributed to "special talents," I modified the Goldman–Widawski/ Elliot–Strenta study design to facilitate comparisons of differences in grades received by students having the same academic major. That is, instead of aggregating all pairwise differences between grades received by students in different academic fields, these differences were segregated according to the academic major of the students who received the grades.

Transcripts of students enrolled at Duke University between the fall of 1995 and the fall of 1999—roughly 15,000 students— were used to compute the pairwise differences under this scheme. For each student who had declared a major by the fall of 1999, pairwise differences between grades received in major courses and nonmajor courses were recorded. These pairwise differences were then averaged to obtain the mean difference in the grades received by students in their major field and all other fields. These differences are displayed in Table 4. Majors with fewer than 30 students were excluded from the analysis,[2] as were mean pairwise differences based on fewer than 10 observations.

The entry in the ith row and jth column in Table 4 represents the average difference, for a student who majored in field i, between grades received in courses in field i and courses in field j. For example, the first entry in Table 4, 0.42, indicates that the average grade received by an Art major in Art courses was 0.42 units higher than the average grade that an Art major received in Biological

---

[2]Only students' primary majors, as reflected in transcript data provided by the Duke University Registrar, were counted toward this total.

TABLE 4

Mean differences in grades received by students in their majors (indicated by the row labels) with courses taken in other departments. Departmental abbreviations are ART = Art, BAA = Biological Anthropology and Anatomy, BIO = Biology, CA = Cultural Anthropology, CHM = Chemistry, CPS = Computer Science, CS = Classical Studies, DRA = Drama, ECO = Economics, ENG = English, FR = French, GEO = Geology, HST = History, LIT = Literature, MTH = Math, MUS = Music, PHL = Philosophy, PHY = Physics, PPS = Public Policy, PS = Political Science, PSY = Psychology, REL = Religion, SOC = Sociology, SP = Spanish.

	ART	BAA	BIO	CA	CHM	CPS	CS	DRA	ECO	ENG	FR	GEO	HST	LIT	MTH	MUS	PHL	PHY	PPS	PS	PSY	REL	SOC	SP	Avg
ART		.42	.53	.13	.93	.20	-.05	-.14	.69	.18	.37	.65	.34	-.12	1.06	-.07	.33	.81	.20	.46	.45	.15	.10	.18	.34
BAA	.22		.98	.07	.93	.27	-.15	-.19	.61	.06	.29	.03	.18	-.03	.89	-.15	.14	.98	-.05	.43	.28	.01	.00	.12	.26
BIO	-.22	-.43		-.40	.19	-.39	-.47	-.56	-.04	-.24	-.27	-.41	-.43	-.43	.18	-.44	-.18	.26	-.26	-.07	-.21	-.35	-.35	-.38	-.25
CA	.33	.52	1.09		1.40	.85	.01	-.01	.93	.22	.37	.55	.22	.19	1.49	-.13	.28	1.01	.35	.41	.41	.05	.56	.30	.50
CHM	-.23	-.18	.29	-.25		-.31	-.27	-.52	-.05	-.14	-.19	-.12	-.08	-.47	.06	-.33	-.12	.18	-.30	-.02	-.13	-.22	-.32	-.35	-.18
CPS	-.05	-.05	.48	-.04	.49		-.09	-.03	.21	.06	.06	.20	.12	-.12	.44	-.12	.04	.37	-.12	.02	.17	-.16	-.19	-.09	.07
CS	.29	.48	.74	.86	1.17	.26		.26	1.00	.16	.61	.23	.46	.08	.97	-.32	.14	.48		.62	.29	.32	.24	.45	.45
DRA	.36	.60	.67	.48	.80	.81			.65	.26	.50	1.36	.49	.07	.85	.02	.57	.81	.62	.76	.69	.16	.47	.39	.53
ECO	-.15	-.09	.41	-.34	.37	-.10	-.33	-.50		-.18	-.10	-.11	-.13	-.44	.33	-.40	-.10	.21	-.06	.02	-.01	-.30	.02	.03	-.11
ENG	-.00	.38	.68	-.07	.92	.18	-.03	-.18	.50		.31	.39	-.10	-.10	.89	-.08	.13	.81	.10	.32	.28	-.09	.15	-.29	.24
FR	-.02	.52	1.16	-.12	1.00	.36	-.15	-.33	.37	.31		.27	.03	-.07	.83	.00	.33	.95	.28	.14	.45	.09	.08	-.26	.26
GEO	.05	.25	.61	.08	.98	.18	.14	-.09	.66	.08	.61		.17	-.25	.96	-.12	.22	.85	.14	.56	.55	.13	.15	.47	.33
HST	-.01	.30	.63	-.04	.77	.38	-.10	-.32	.51	.18	.25	.39		-.02	.92	-.20	.19	.74	.08	.25	.38	-.07	.08	.08	.23
LIT	.14	.87	.86	-.04	1.21	.54	-.23	-.21	.73	.06	.41	.72	.32		.99	-.02	.27	1.07	.06	.54	.52	.17	.07	.16	.40
MTH	-.08	-.07	.33	-.19	.16	-.06	-.07	-.35	-.04	.10	.29	-.24	-.13	-.18		-.37	-.07	.18	-.23	.17	-.07	-.22	.02	-.09	-.09
MUS	.00	.52	1.09	.21	.79	.14	-.26	-.14	.75	-.16	.19	.24	-.20	-.15	.63		.35	.73	.36	.31	.48	.18	.16	.40	.33
PHL	.02	.30	.68	.03	.75	.25	.14	-.01	.34	.38	.16	.36	-.07	-.05	.62	-.28		.49	-.15	.22	.41	-.07	-.06	.03	.18
PHY	-.35	.55	.36	-.16	.22	-.06	-.23	-.46	-.10	.06	.05	.16	-.01	-.52	.05	-.40	-.20		-.34	-.72	-.35	-.39	-.10	-.32	-.17
PPS	-.09	.10	.38	-.21	.65	-.03	-.22	-.35	.32	-.29	.07	.22	-.00	-.24	.71	-.23	.01	.53		.10	.20	-.15	-.10	-.07	.06
PS	-.08	.15	.36	-.22	.66	.21	-.24	-.31	.45	-.07	.19	.19	-.04	-.30	.73	-.30	.02	.54	.02		.25	-.13	-.15	-.07	.08
PSY	-.03	.07	.76	-.08	.77	.07	-.15	-.29	.37	-.01	.16	.19	.08	-.14	.73	-.21	.10	.72	.00	.25		-.13	-.12	-.10	.13
REL	.13	.17	.53	.04	.46	.10	-.12	.15	.47	.06	.30	.30	.10	-.36	.75	-.02	.27	.67	.10	.22	.38		.16	.15	.21
SOC	.15	.37	.79	.15	1.17	.50	-.09	-.37	.66	.03	.35	.47	.14	-.38	.96	-.39	.11	1.05	-.01	.38	.55	-.06		.09	.29
SP	.03	.18	.67	.01	.78	.42	-.09	-.08	.84	.16	-.01	1.24	.03	-.07	.78	-.04	.25	.80	.19	.36	.45	.01	.19		.31
Mean	.02	.26	.66	-.00	.76	.21	-.12	-.24	.47	.04	.20	.29	.10	-.18	.73	-.20	.13	.66	.05	.25	.28	-.05	.02	.01	.18

Anthropology and Anatomy courses. This value is approximately equal to the difference between an A− and B+.

The row means in Table 4 list the average difference between grades received by students in their major and the grades they received in nonmajor courses. From the row sums, it is apparent that biology, chemistry, economics, mathematics, and physics majors received, on average, higher grades for courses taken outside of their majors than for courses taken in their majors, while all other students received higher average grades in their major courses. The average difference between grades received "in major" and "out of major" for all students was 0.18, about one-half the value of a "+" or "−" grade in Duke's grading system. This positive difference should be expected and can be attributed to increased student interest and aptitude in major classes. Furthermore, students usually take more small upper-level classes in their major than they do outside of their major, and small upper-level classes tend to be graded more leniently than large introductory classes.

Differences between row means in Table 4 can be attributed to either disparities in grading policies between fields, or to disparities in the amount of "special talent" required by majors in different fields. Like the Goldman–Widawski and Strenta–Elliot studies, the two explanations are largely confounded in the row effects.

The column means in Table 4 represent the average difference in grades received by nonmajors in courses in their field of study and grades received in the given field. For example, the entry of 0.76 at the bottom of the chemistry column indicates that nonchemistry majors received grades that were, on average, 0.76 units higher in courses in their major than in chemistry. Similarly, the value of −0.24 in the drama column indicates that nondrama students received higher grades in drama than they did in their own major.

In contrast to differences in the row means, differences in the values of the column means are hard to explain using the notion of "special talents." For while it is quite plausible that drama majors might have special abilities in drama that allow them to obtain higher grades in drama classes, it is substantially less plausible that students from every other academic major who happened to take drama classes also possessed the same special abilities. Yet students from every other major received higher average grades in drama than they did for courses taken in their own majors.

The average value of the column means in Table 4 is, like the row means, 0.18. Values less than 0.18 thus suggest more lenient than average grading practices; values greater than 0.18 suggest more stringent than average grading practices.

Differences in the column means of Table 4 provide an estimate of differences in grading practices between fields. For example, the difference between grading practices in chemistry and drama courses is $0.76 - (-0.24) = 1.0$, or an entire letter grade average difference. A graphical depiction of the column means is provided in Figure 1. Pairwise differences in grades computed under the protocol used by Goldman and Widawski and Elliot and Strenta are listed as an appendix to this chapter for further comparison.

The analyses summarized in Table 4 and the studies of Goldman–Widawski and Elliot–Strenta provide nearly incontrovertible evidence of systematic disparities in grading practices between fields. Furthermore, the magnitudes of the pairwise differences obtained over a nearly thirty-year period at both public and private institutions is remarkably consistent with regard to both academic field and ability levels of students. Folklore involving disparities in grading between academic divisions has a basis in fact: Natural sciences and mathematics departments do tend to grade more stringently than social science departments, and social science departments do tend to grade more stringently than humanities departments.

METHODS FOR GRADE ADJUSTMENT

A primary goal of the Goldman and Widawski [GW76], Strenta and Elliot [SE87], and Elliot and Strenta [ES88] studies was to explain the relative inability of SAT scores and high school GPA to predict the success of college students. It was the view of each of these authors that the failure of SAT scores and high school GPA to accurately predict college success could be attributed, at least in part, to deficiencies in the outcome measure, college GPA.

FIGURE 1

Illustration of the average difference in the grades students received in courses in their major and outside of their major. Specific values depicted along the axis correspond to the column means listed in Table 4.

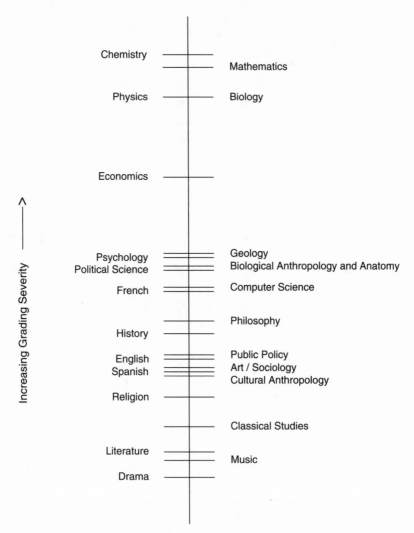

Larkey and Caulkin [LC92] and Caulkin, Larkey, and Wei [CLW96] took this idea a step further by incorporating the notion of pairwise differences computed in previous studies into a more formal linear regression model. Their goal in so doing was to provide a simple method for adjusting individual student GPAs for disparities in grading practices across disciplines and instructors.

In their regression models, Larkey and colleagues assumed that the grade of the ith student in the jth class, denoted by $Y_{i,j}$, resulted from the combination of a student effect, denoted by g_i, a course grading effect, denoted by c_j, and a random error, $e_{i,j}$. More specifically, they assumed that

$$Y_{i,j} = g_i + c_j + e_{i,j}. \tag{7.1}$$

In this equation, the course difficulty effect, c_j, represents something akin to the pairwise differences examined in the previous section, except that a distinct value is estimated for every course. The student effect, g_i, can be interpreted as representing student i's "true" GPA, which, in theory, could be directly observed if there were no differences in the way courses were graded and each student took an arbitrarily large number of classes.[3]

Larkey, Caulkin, and Wei fit variations of this model to grades obtained by students at Carnegie Mellon University in order to demonstrate both the feasibility of fitting the model to actual grade data and to emphasize the potential importance of adjusting student grades. In order to illustrate their method and to facilitate comparisons with another grade adjustment method, this regression model was applied to the grades of all full-time Duke University students who matriculated at Duke in the autumn of 1995. The transcripts of two of these students have been summarized in Tables 5 and 6. In addition to the courses that these students took before the fall of 1999 and the grades they received in these courses, these tables also contain the mean grade assigned to all students in each course taken by these students, the percentage of students in each class who received the same grade or a higher grade than the given student,

---

[3]The random error, $e_{i,j}$, was implicitly assumed to be normally distributed with zero mean.

TABLE 5

Sample transcript and associated adjustments using additive grade adjustment
model of Larkey and Caulkin [LC92] and Caulkin, Larkey, and Wei [CLW96].

GPA: 3.38   GPA based rank: 721
Additive GPA: 3.19   Additive GPA rank: 960

Course	Grade	Mean Grade	% Same Grade	% Higher Grade	Mean Add-GPA	Adjusted Grade
GER 001	A–	3.85	0.50	0.50	3.31	3.10
MTH 031L	D–	2.54	0.04	0.96	3.22	1.63
PSY 011	D+	2.53	0.04	0.83	3.27	1.93
UWC 005	B+	3.54	0.10	0.80	3.29	3.03
DRA 101S	A–	3.71	0.42	0.42	3.34	3.28
GER 002	A–	3.85	0.50	0.50	3.31	3.10
PHL 044S	B	3.09	0.31	0.45	3.41	3.28
STA 010D	B	3.24	0.26	0.56	3.26	2.92
DRA 196S	B+	3.72	0.40	0.60	3.43	2.97
GER 065	B	3.18	0.23	0.62	3.19	2.69
PPS 180S	A	3.46	0.40	0.00	3.24	3.72
PS 091D	A–	3.08	0.15	0.12	3.23	3.77
DRA 140S	B+	3.50	0.50	0.50	3.16	2.90
GER 066	B	3.34	0.13	0.80	3.25	2.62
PS 120	B	3.00	0.47	0.26	3.33	3.28
PS 131	B+	3.15	0.32	0.27	3.41	3.45
PS 100E	B+	3.68	0.23	0.77	3.29	2.87
GER 117S	B+	3.33	0.43	0.50	3.26	2.93
HST 103	A–	3.60	0.31	0.34	3.23	3.28
HST 160	B+	3.57	0.36	0.56	3.37	3.06
PPS 181S	A	3.46	0.40	0.00	3.40	3.90
PS 189	B+	2.77	0.50	0.17	3.36	3.29
EDU 100	A	3.51	0.27	0.00	3.21	3.64
HST 092X	A	3.79	0.50	0.00	3.41	3.58
HST 153	A	3.60	0.44	0.00	3.07	3.41
PS 187S	B+	3.01	0.37	0.25	3.15	3.18
PPS 082	A	2.55	0.19	0.00	3.08	3.36
PPS 195	A–	3.40	0.20	0.10	3.28	3.54
PPS 195S	A	3.32	0.25	0.00	3.20	3.53
PS 203S	B+	3.31	0.27	0.36	3.15	3.08
DRA 136S	B+	3.70	0.25	0.75	3.17	2.72
HST 196S	A	3.98	0.89	0.06	3.41	3.39
PPS 176S	B	3.28	0.31	0.54	3.33	3.00
PPS 195S	A	3.39	0.37	0.00	3.18	3.74
	3.38	3.35	0.33	0.37	3.27	

TABLE 6

Sample transcript and associated adjustments using additive grade adjustment
model of Larkey and Caulkin [LC92] and Caulkin, Larkey, and Wei [CLW96].

```
GPA: 3.37      GPA based rank: 733
Additive GPA: 3.64  Additive GPA rank: 322
```

Course	Grade	Mean Grade	% Same Grade	% Higher Grade	Mean Add-GPA	Adjusted Grade
BIO 093S	B+	3.53	0.39	0.50	3.49	3.24
CA 060S	B+	3.12	0.31	0.37	3.28	3.39
CHM 011L	B+	2.87	0.15	0.28	3.31	3.66
UWC 007	A	3.79	0.45	0.09	3.42	3.60
ART 070D	A	3.37	0.30	0.01	3.40	3.95
CHM 012L	B-	2.90	0.12	0.60	3.37	3.14
MTH 041	B+	2.64	0.17	0.17	3.37	4.00
PSY 092	A	3.23	0.32	0.00	3.39	4.00
CHM 151L	C	2.46	0.18	0.61	3.44	2.84
ECO 052	C+	3.26	0.10	0.88	3.41	2.43
PS 137D	B+	3.25	0.26	0.29	3.30	3.30
SP 014	A+	3.55	0.12	0.00	3.06	3.44
BIO 118	B	2.65	0.15	0.35	3.52	3.77
CHM 152L	B	2.60	0.21	0.31	3.51	3.82
MUS 125	A-	3.26	0.21	0.17	3.33	3.63
PSY 119A	A+	3.44	0.15	0.00	3.33	3.85
BIO 119	A-	2.86	0.07	0.20	3.54	4.35
PHY 053L	B+	2.97	0.12	0.31	3.50	3.77
PSY 091	A	2.90	0.23	0.00	3.44	4.40
PSY 099	A	3.13	0.21	0.00	3.24	4.04
PSY 137	B	3.01	0.18	0.46	3.38	3.34
BIO 151L	B+	2.87	0.29	0.13	3.60	3.96
PHY 054L	B-	3.01	0.09	0.63	3.53	3.19
PSY 162S	B	3.19	0.50	0.43	3.41	3.18
STA 110E	A	2.95	0.21	0.07	3.39	4.30
	3.37	3.07	0.22	0.27	3.40	

the mean adjusted GPA of other students enrolled in the class, and
the adjusted grade of the student in the class. The adjusted grade
was computed as the difference in the received grade $(Y_{i,j})$ and the
estimated course effect variable (c_j). Note that the two students
chosen for this illustration have approximately the same GPA;

the GPA of the student whose transcript is listed in Table 5 was 3.38, while the GPA of the student whose transcript is displayed in Table 6 was 3.37. Based on their unadjusted GPAs, both students rank near the middle of their class, which included about 1,400 students.[4]

Despite the similarity of their GPAs, the academic records and achievement of these students is markedly different. For example, the mean course grade assigned in the courses that the first student took was 3.35, while the mean course grade in classes taken by the second student was 3.07. So the first student performed at about an average level in his classes, while the second did substantially better than his classmates. This observation is further supported by examining the average proportion of students who received the same or higher grades in the students' classes. The first student received a lower grade, on average, than about a third (37%) of his classmates in the classes he took, and received the same grade as the other students another 33% of the time. The first student was thus solidly in the middle range of achievement when compared to his peers.

In contrast, the second student received a lower grade, on average, than about one-quarter of his classmates (27%), and received the same grade as his peers about 22% of the time. He outperformed 51% of his classmates, on average, in the courses he took, as compared to 30% for the first student.

But even these numbers do not provide an accurate comparison of the relative performance of these students, because the quality of their classmates also differed. This fact is reflected by the column in the table labeled "Mean Add-GPA," which lists the mean adjusted GPA of students in each class, where GPA has been adjusted by the additive regression model. While the first student took classes with other students who, on average, had additively adjusted GPAs of 3.27, the second student took classes with students who had an average additively adjusted GPA of 3.40. Thus, the second student seems to have taken classes with better students. Overall, it seems

---

[4]Only the grade records of students who had completed 25 graded courses by the spring of 1999 and who had matriculated after the fall of 1994 were used to compute these summaries.

clear that the second student did significantly better in his classes than did the first, yet their GPAs are essentially the same.

In addition to exposing differences in the individual performance of these students, the entries in these tables shed further light on the magnitude of the disparities between grading practices in the humanities and natural sciences. For example, the student transcript listed in Table 5 indicates that the first student took two introductory German classes, GER 001 and GER 002, and received an A− in each. In both classes, however, an A− turned out to be the lowest grade awarded to any of the students included in this cohort. Grades of A or A− were also awarded to this student in four other classes in which his performance was average or below average: An introductory drama class (DRA 101S) and three history classes (HST 103, HST 092X, HST 196S). That is not to say, however, that the second student did not also occasionally benefit from lenient grading, only that he benefited less; the only A he got for "average" performance was in a required writing course, UWC 007.

The final column in Tables 5 and 6 illustrates the corrections made to each course grade by the additive model specified in (7.1). For example, the first course listed for the student in Table 5, an A− in introductory German, is assigned a value of 3.7 in normal GPA computations. However, the grade distribution in this course was deemed by the model to be more lenient than average, and so the course effect variable was estimated as 0.6. As a consequence, all grades awarded to students in this class were effectively decreased by 0.6 grade units when their "true" GPA was calculated. Conversely, the D− assigned to this student in the second class, an introductory calculus course, was incremented in the model from a value of 1.0 to 1.63.

Note that the "Additive GPA" listed at the top of each table does not represent precisely the "true GPA" (g_i) presented in (7.1). The reason for this is that the corrected GPA proposed in [LC92] and [CLW96] is not guaranteed to provide GPAs that are less than 4.0. This irregularity is caused by upward adjustment of A's awarded in stringently graded classes. Because GPAs are traditionally scaled to fall in the range 0–4.0, the adjusted GPAs from the additive model were rescaled to have the same distribution

216

CHAPTER 7

as the unadjusted GPAs. Rescaling was performed by assigning the highest unadjusted GPA to the student who had the highest value of g_i, the second-highest GPA to the student who had the second-highest value of g_i, and so on. An important benefit—or detriment, depending on your perspective—of this transformation is that it exactly maintains the mean GPA reported for all students and so has no direct impact on "grade inflation."

The consequences of grade adjustments for the students whose transcripts are listed in Tables 5 and 6 are potentially quite substantial. The GPA of the first student, whose classes were graded more leniently than average, decreased by 0.19 units, moving him from the middle of his class to approximately the bottom quartile. The second student, whose classes were graded substantially more stringently than average, moved in the opposite direction, landing him in the top quartile of his class. Had this student graduated with a 3.64 GPA rather than a 3.37 GPA, his chances for obtaining a fellowship to a prestigious graduate school or admission to a professional school would likely have more than doubled.

The potential consequences of GPA adjustment for the broader population of undergraduate students are equally serious. For this same cohort of Duke students, Figure 2 demonstrates the changes that would occur in the rankings of students within this class if ranks were based on adjusted GPAs rather than standard GPA. For students at the top or bottom of their class, the changes tend to be comparatively small: Cream does rise to the top (and vice versa!). For students in the middle half of the class, though, the differences can be dramatic, and changes to GPA and career options for many of these students would mirror the effects reported for the two students considered above.

Attempts to reform undergraduate grading through the introduction of adjustment schemes like those proposed in [LC92, CLW96] can be politically explosive, and any scheme proposed is likely to be subjected to close, and often cynical, scrutiny. It is therefore critical to identify both the strengths and weaknesses of an adjustment strategy before it is proposed for implementation. In this regard, the additive model proposed by Larkey, Caulkin,

FIGURE 2

Comparison of student ranks based on standard GPA and adjusted GPA.

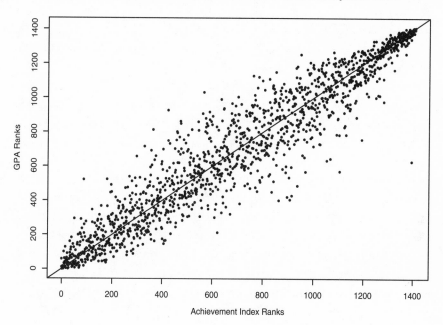

and Wei has several significant strengths, but it also has substantial flaws.

Among its strengths are its ease of specification, computational simplicity, and relative conceptual transparency. The regression model that lies at the heart of this adjustment scheme can be specified by a single linear equation. Computationally, the model is trivial to implement using the algorithm specified in [CLW96]. Also, the linear regression model upon which it is based is perhaps the most commonly used statistical model in the social and natural sciences, and is undoubtedly familiar to anyone who has taken an introductory statistics course.

There is, however, a fundamental deficiency in this model. It involves the fact that an additive adjustment is applied to all of the grades assigned in every class.

To understand the negative aspects of making a single additive adjustment to all grades assigned in a particular course, consider two students who, with one exception, take identical courses for four years, and receive A+'s in all of their courses. In their fourth year, the first student decides to take, say, a drama course that the second student doesn't. The instructor of the drama course awards every student in his course an A+, which means that the first student's perfect record remains intact.

But because the drama instructor adopted a lenient grading policy in assigning all A+'s, the adjusted value of the A+ in this class is likely to be comparatively low. This means that the adjusted GPA of the first student is decreased as a consequence of receiving an A+ in the drama class. This makes his adjusted GPA *lower* than the adjusted GPA of the second student, despite the fact that both students received the highest possible grade in all of their courses, and the fact that the first student took one more course. In practice, exactly this problem manifests itself on a fairly regular basis for students who receive the highest marks in humanities classes. For example, the student transcript displayed in Table 6 shows that this student received an A+ in an introductory Spanish course and an A in a psychology course, PSY 091. The A+, by virtue of the fact that the mean grade in the Spanish course was 3.55 and that 12% of the class received an A+, was adjusted downward to a value of 3.44. In contrast, the A in PSY 091, which was awarded in a class that had a 2.90 mean grade, was adjusted upward to a value of 4.40, nearly an entire letter grade higher than the value of the adjusted A+ in Spanish.

The resolution of this dilemma is problematic. On the one hand, an A awarded in a class that had a B average should be valued more highly than an A awarded in a class that had an A− average. On the other, it seems unfair to penalize a student who truly performed at an outstanding level simply because the student's instructor graded leniently. Unfortunately, the grade record provides no indication of which A's in a leniently graded class truly represented outstanding achievement, and which did not.

In 1997, I proposed a more complicated adjustment scheme, called the achievement index, to overcome this and several more subtle deficiencies of the additive model for grade adjustment [Joh97]. Like the additive adjustment scheme described above, this scheme also estimates a "true" GPA for each student, but the connection between student achievement and received grades is modeled through a more sophisticated, nonlinear statistical model called a multirater–ordinal probit model (e.g., [Joh99]). Although the details of the statistical model underlying this adjustment scheme are too involved to describe here (a heuristic explanation of the adjustment scheme is described in this chapter's appendix), its key feature is that it takes account of the grade thresholds that instructors use in assigning all of the grades in their classes. That is, the range of student achievement levels that an instructor classified as falling into the A range is estimated for each class, as is the range of achievement classified as an A−, as is the range of achievement associated with a B+, and so on. As a consequence of this feature, students who receive, say, A's from instructors who assign nothing but A's are neither penalized nor rewarded, since the associated range of achievement associated with an A assigned by such an instructor spans the entire range of student achievement.

Adjustments obtained using the achievement indices are quite similar to those using the additive adjustments proposed by Larkey, Caulkin, and Wei. To illustrate these similarities, results obtained by calculating adjusted grades using the achievement index method for the two students whose transcripts were considered under the Larkey–Caulkin–Wei additive model are also listed in Tables 7 and 8.

Many of the comments made concerning the summaries listed in Tables 5 and 6 apply also to Tables 7 and 8. The primary difference between these tables, and the models that underlie them, manifests itself in the final column. In the case of the additive model, this column contained a single adjusted grade value. In Tables 7 and 8, an interval representing the estimated range of performances that correspond to the grade assigned in the second column is presented.

To understand the meaning of this interval, consider the first row of Table 7, in which the student received a grade of A− in an introductory German course. As noted previously, an A− was the lowest grade assigned in this class, which implies that a grade of A− covers the entire range of student performance below the range estimated for an A. Thus, the range of performance associated with an A− in

TABLE 7

Results from the application of the multirater–ordinal probit model to student grades. The second column in the table lists the grade received by the student in each of the courses shown in the first column. The third column provides the mean course grades for each course. The fourth and fifth columns list the proportion of students in each class that received the same or higher grade than the given student. The sixth column lists the mean adjusted GPA for students in the course, adjusted using the multirater–ordinal probit model. The final column provides the estimated grade thresholds for the grades received. These grade thresholds represent rescaled versions of the thresholds displayed in Figure 3; they were rescaled so as to fall in the interval 0–4.

```
GPA: 3.38     GPA based rank: 721
AI-GPA: 3.20  Achievement Index rank: 947
```

Course	Grade	Mean Grade	% Same Grade	% Higher Grade	Mean AI-GPA	Grade Cutoffs
GER 001	A−	3.85	0.50	0.50	3.38	(0.00, 3.25)
MTH 031L	D−	2.54	0.04	0.96	3.24	(0.00, 1.97)
PSY 011	D+	2.53	0.04	0.83	3.28	(1.97, 2.23)
UWC 005	B+	3.54	0.10	0.80	3.30	(2.11, 2.58)
DRA 101S	A−	3.71	0.42	0.42	3.40	(2.60, 3.42)
GER 002	A−	3.85	0.50	0.50	3.38	(0.00, 3.25)
PHL 044S	B	3.09	0.31	0.45	3.41	(2.89, 3.42)
STA 010D	B	3.24	0.26	0.56	3.29	(2.46, 3.06)
DRA 196S	B+	3.72	0.40	0.60	3.49	(0.00, 3.26)
GER 065	B	3.18	0.23	0.62	3.19	(2.07, 2.69)
PPS 180S	A	3.46	0.40	0.00	3.28	(3.34, 4.00)
PS 091D	A−	3.08	0.15	0.12	3.22	(3.50, 3.81)
DRA 140S	B+	3.50	0.50	0.50	3.18	(0.00, 3.02)
GER 066	B	3.34	0.13	0.80	3.23	(1.00, 2.28)
PS 120	B	3.00	0.47	0.26	3.38	(3.17, 3.41)
PS 131	B+	3.15	0.32	0.27	3.45	(3.23, 3.69)
PS 100E	B+	3.68	0.23	0.77	3.31	(0.00, 2.63)
GER 117S	B+	3.33	0.43	0.50	3.25	(1.00, 3.02)
HST 103	A−	3.60	0.31	0.34	3.22	(2.81, 3.36)
HST 160	B+	3.57	0.36	0.56	3.41	(2.37, 3.21)
PPS 181S	A	3.46	0.40	0.00	3.47	(3.56, 4.00)
PS 189	B+	2.77	0.50	0.17	3.38	(2.74, 3.69)
EDU 100	A	3.51	0.27	0.00	3.19	(3.62, 4.00)
HST 092X	A	3.79	0.50	0.00	3.42	(3.34, 4.00)
HST 153	A	3.60	0.44	0.00	3.09	(3.04, 4.00)
PS 187S	B+	3.01	0.37	0.25	3.14	(2.73, 3.36)
PPS 082	A	2.55	0.19	0.00	3.04	(3.42, 4.00)
PPS 195	A−	3.40	0.20	0.10	3.25	(3.48, 3.84)
PPS 195S	A	3.32	0.25	0.00	3.18	(3.45, 4.00)
PS 203S	B+	3.31	0.27	0.36	3.18	(2.82, 3.24)
DRA 136S	B+	3.70	0.25	0.75	3.09	(0.00, 2.42)
HST 196S	A	3.98	0.89	0.06	3.44	(2.08, 3.99)
PPS 176S	B	3.28	0.31	0.54	3.29	(2.38, 3.05)
PPS 195S	A	3.39	0.37	0.00	3.11	(3.26, 4.00)
	3.38	3.35	0.33	0.37	3.28	

TABLE 8

Similar to Table 7, but for a student who took more stringently graded courses.

GPA: 3.37   GPA based rank: 733
AI-GPA: 3.61   Achievement Index rank: 378

Course	Grade	Mean Grade	% Same Grade	% Higher Grade	Mean AI-GPA	Grade Cutoffs
BIO 093S	B+	3.53	0.39	0.50	3.51	(2.71, 3.44)
CA 060S	B+	3.12	0.31	0.37	3.31	(2.81, 3.51)
CHM 011L	B+	2.87	0.15	0.28	3.31	(3.31, 3.56)
UWC 007	A	3.79	0.45	0.09	3.43	(3.31, 4.00)
ART 070D	A	3.37	0.30	0.01	3.44	(3.61, 4.00)
CHM 012L	B-	2.90	0.12	0.60	3.37	(2.92, 3.12)
MTH 041	B+	2.64	0.17	0.17	3.38	(3.52, 3.87)
PSY 092	A	3.23	0.32	0.00	3.42	(3.60, 4.00)
CHM 151L	C	2.46	0.18	0.61	3.44	(2.43, 2.67)
ECO 052	C+	3.26	0.10	0.88	3.41	(1.63, 2.44)
PS 137D	B+	3.25	0.26	0.29	3.33	(3.16, 3.53)
SP 014	A+	3.55	0.12	0.00	3.08	(3.69, 4.00)
BIO 118	B	2.65	0.15	0.35	3.50	(3.17, 3.41)
CHM 152L	B	2.60	0.21	0.31	3.49	(3.39, 3.64)
MUS 125	A-	3.26	0.21	0.17	3.31	(3.46, 3.89)
PSY 119A	A+	3.44	0.15	0.00	3.35	(3.85, 4.00)
BIO 119	A-	2.86	0.07	0.20	3.49	(3.73, 3.83)
PHY 053L	B+	2.97	0.12	0.31	3.46	(3.50, 3.66)
PSY 091	A	2.90	0.23	0.00	3.38	(3.60, 4.00)
PSY 099	A	3.13	0.21	0.00	3.18	(3.53, 4.00)
PSY 137	B	3.01	0.18	0.46	3.35	(3.04, 3.36)
BIO 151L	B+	2.87	0.29	0.13	3.55	(3.57, 3.90)
PHY 054L	B-	3.01	0.09	0.63	3.48	(3.07, 3.23)
PSY 162S	B	3.19	0.50	0.43	3.39	(2.26, 3.41)
STA 110E	A	2.95	0.21	0.07	3.34	(3.52, 3.91)
	3.37	3.07	0.22	0.27	3.39	

this course extends from 3.25 down to 0. In other words, any student who performed at a level of 3.25 or lower (on the standardized scale) would be expected to receive an A– in this course. The range of performance associated with an A was 3.25 to 4.0.

This example illustrates a more general property of the achievement index. Namely, the upper range for the highest grade received in a class is always 4.0, regardless of what that grade is or how many students got it. Consequently, a student can never be penalized for receiving the highest grade in a class using the achievement index adjustment.

Along similar lines, in the hypothetical situation in which an instructor assigns A's to all students, the lower threshold for an A is 0.0 (on the standardized GPA scale) and the upper threshold is a 4.0. Since any level of student achievement is then associated with an A, grades of A are implicitly deemed uninformative and have no effect on a student's GPA.

Superficially, this treatment of uniform classroom grade assignments appears entirely reasonable; an instructor who assigns everyone in a class the same grade has not provided any comparative information regarding the performance of students in that class, and so their grades cannot be used to differentiate between the performance levels of different students. It is interesting to note, however, that it was exactly this aspect of the achievement index adjustment that proved most controversial when it was proposed as an alternative to standard GPA at Duke University in 1996.

The root of this controversy stems from the fact that in practice, when only one grade is assigned in a class, it is inevitably an A or A+. When the achievement index was considered for use as a mechanism to adjust GPAs for students at Duke, instructors who regularly assigned uniformly high grades quickly realized that the achievement index adjustment would make their grades irrelevant in the calculation of student GPAs. Worse still, many students noticed the same thing. To thwart the adoption of the achievement index, these high-grading instructors and their student beneficiaries adopted the position that an A represented an *objective* assessment of student performance. An A was an A was an A. For them, it represented "excellent" performance on some well-defined but unobservable scale. Indeed, by the end of the debate, several literary theorists had finally identified an objective piece of text: a student grade.

The vacuousness of these assertions becomes clear in light of the data presented in the preceding section. For example, consider the repercussions of an instructor who assigns A's to all his students. By so doing, this instructor has refused to rank students within his class according to achievement. Yet, because the mean grade assigned at most universities has not yet reached a 4.0, this instructor has tacitly ranked his students higher than all students who elected not to take his class. The aftermath of this instructor's actions is the elevation of his students' GPAs relative to the GPAs of all students who chose not to take his class. Clearly, this practice is unfair to students who, either by choice or by accident, elect to take courses from instructors who graded more stringently.

How does the accuracy of adjusted grades and GPA compare to unadjusted grades and GPA? This is a difficult question to answer owing to the fact that no gold standard for grading or student learning exists. External measures of student ability, like SAT or high school GPA, do not necessarily measure the same qualities that college grading schemes are intended to measure, and, of course, do not measure these qualities at the time students are attending college. Because objective measures of student performance are not available, indirect measures of the validity of these performance measures must be found. One such criterion can be based on the ability of an index to predict future data. For student grades, this suggests that performance measures can be compared by their ability to predict which of two students will do better in a class that they both take.

This criterion was used in [Joh97] to compare standard GPA, additively adjusted GPA, and achievement-index-adjusted GPA for students who had taken at least two courses together. The reason that the comparison was restricted to students who had taken two courses together is that it permits a baseline error rate to be established. This baseline can be established by examining, in each pair of classes taken by two students, the proportion of times that one student received the higher grade in one course and the lower grade in the other. Clearly, no univariate index of student achievement can account for such reversals.

Applying this criterion to transcript data collected for a class of Duke University undergraduates, the lowest achievable error rate in predicting which of two students would do better in a course that they both elected to take was found to be 0.131. For this same population of student pairs, the additional error associated with standard GPA was estimated to be 0.075, which represents a 57% increase over the ideal rate. The increase associated with the additively adjusted GPA was 0.043, or an increase of 33% over the best possible rate. The corresponding value for the achievement-index-adjusted GPA was 0.037, a 28% increase over the ideal rate. The error associated with the achievement index was thus 50% lower than with standard GPA.

In interpreting these predictions, it is important to note that most of the comparisons made involved students with similar majors, since students with different majors were unlikely to take two courses together. Because students in the same major tend to take courses that are graded more comparably than students with different majors, the error estimates obtained for standard GPA are probably conservative in the sense that the errors associated with the standard GPA would be larger across the broader population of students.

As an additional check on the accuracy of the adjustments to GPA, the proportion of the variance that could be explained in predicting college GPA using high school GPA and SAT scores was computed. High school GPA and SAT scores were able to account for 25.2% of the variation in the standard GPA of students. For additively adjusted GPAs, this proportion increased to 33.8%, and for achievement-index-adjusted GPAs to 34.6%. These figures imply that the achievement-index-adjusted GPA correlates best with external measures of student ability like SAT and high school GPA, although, for reasons cited above, the higher correlation of the achievement-index adjustment with these external measures of student ability may not be a clear indication that this adjustment method should be preferred to the additive scheme.

DISCUSSION

From the initial regression studies of Goldman, Schmidt, Hewitt, and Fisher [GSHF74] through the analysis of recent Duke University transcript data, a consistent pattern in the way various academic disciplines distribute academic capital has emerged. Generally speaking, humanities and most social science departments have adopted grading policies that provide comparatively lucrative reward structures for student performance, while natural sciences, mathematics, and economics departments have adopted more miserly distribution schemes.

The fact that grading practices differ across disciplines is often challenged on the grounds that observed differences in grade distributions might be attributed to differences in special aptitudes of students. Unusually high grades in, for example, an art class are often justified on the premise that students in the art class were especially "artistic."

The analyses of the first section of this chapter render this argument untenable. Drama classes at Duke provide a case in point; literature classes provide another. With the exception of students who majored in cultural anthropology, classical studies, and drama, which were themselves among the most leniently grading departments, students from all other majors received higher grades in literature than they did in courses in their own majors. For the "special talents" argument to explain this phenomenon, it is necessary to assume that literature majors not only have special talents in literature that make their performance in literature unusually outstanding, and that these special talents exceed the special talents of, say, economics majors in economics, but that students from essentially all other majors who elected to take literature courses had these same special talents. At this point, the argument that apparent grading disparities are due to special talents of students passes from merely implausible to completely absurd. The more reasonable explanation is that literature courses were simply graded more leniently than courses in most other disciplines.

Given that grading practices do differ between courses and disciplines, and that such differences cause undesirable distortions in a number of educational processes, including student enrollment decisions and student and faculty assessment, the question now becomes how to remedy these differences. Perhaps the most innocuous scheme would involve the use of grade adjustment methodology to provide faculty with estimates of the standardized grade intervals that correspond to the grades awarded in their classes, and to encourage them to adjust their grading policies so as to more closely match the norm. The success of such a scheme would depend not only on the willingness of humanities faculty to tighten their grading policies, but also the cooperation of many science faculty in loosening theirs.

More controversial strategies for adjusting grades to account for disparate grading policies include displaying adjusted grades or GPAs on student transcripts and the use of adjusted GPAs to modify university-imposed grading standards. More specific discussion of the latter point is delayed until the next chapter. The former—correcting student grades for differences in instructor grading practices—provides administrators with a tool to correct student grades for grading inequities without explicitly requiring changes in faculty grading practices. Although correcting student grades for disparities in instructor grading practices would not necessarily eliminate the impact of disparate grading practices on student evaluations of teaching, and might not reverse the trend for students to enroll disproportionately in classes that were graded leniently, it would at least make student evaluation more equitable. And over time, such adjustments might prompt better students to apply pressure on their instructors to differentiate in their grading practices.

Appendix

Pairwise Differences in Grades at Duke University (a là Goldman and Widawski [GW76])

See Table 9 on page 227.

Explanation of Achievement Index Adjustment Scheme

A heuristic explanation of what the achievement index is and how it is computed follows. For each student, say student i, let a_i denote the average level of academic achievement exhibited by that student

TABLE 9

Mean differences in grades received by students taken in two departments using the design of Goldman and Widawski (1976). Departmental abbreviations are ART = Art, BAA = Biological Anthropology and Anatomy, BIO = Biology, CA = Cultural Anthropology, CHM = Chemistry, CPS = Computer Science, CS = Classical Studies, DRA = Drama, ECO = Economics, ENG = English, FR = French, GEO = Geology, HST = History, LIT = Literature, MTH = Math, MUS = Music, PHL = Philosophy, PHY = Physics, PPS = Public Policy, PS = Political Science, PSY = Psychology, REL = Religion, SOC = Sociology, SP = Spanish.

	ART	BAA	BIO	CA	CHM	CPS	CS	DRA	ECO	ENG	FR	GEO	HST	LIT	MTH	MUS	PHL	PHY	PPS	PS	PSY	REL	SOC	Avg
ART																								.14
BAA	-.19																							-.00
BIO	-.34	.59																						-.38
CA	.00	.27	-.59																					.23
CHM	-.55	-.78	-.07	-.79																				-.54
CPS	-.07	-.04	.44	-.22	.04																			-.01
CS	.08	.27	.54	.20	-.22	.82																		.24
DRA	.20	.49	.71	.04	.52	.26	.15																	.40
ECO	-.31	-.22	.17	-.09	-.17	.61	-.44	-.13																-.24
ENG	-.04	.22	.44	.39	-.27	.53	.08	-.42	-.20															.13
FR	-.23	-.01	.39	-.02	-.26	.76	-.03	-.15	-.46	.20														.06
GEO	-.22	-.02	.45	.30	-.01	.54	-.01	-.60	.30	.30	-.17													.01
HST	-.11	.17	.26	.20	.00	.76	.17	.00	.18	.10	.10	-.16												-.08
LIT	.09	.48	.75	.53	.80	.01	.37	.53	.37	.18	.27	.53	.17											.30
MTH	-.59	-.68	-.37		.01	-.74		-.86	-.66	-.54	-.77	-.63	-.82	-.70										-.55
MUS	.11	.36	.79		.67	-.38			.18	.24	.36		.27	.05	.61									.30
PHL	-.11	.06	.35		.46	.21			-.08	.04	-.00		.19	.14	.12	-.08								.01
PHY	-.60	-.81	.18		-.05	-.64			-.58	-.54	-.56		-.66	-.67	-.77	-.08	-.61							-.51
PPS	-.09	.12	-.26		.59	-.00			-.06	.13	-.55		-.15	-.08	.13	.13	-.08							.05
PS	-.19	.02	.18		.44	-.22			-.18	-.24	.01		.12	-.08	-.15	-.23	-.24							-.05
PSY	-.12	-.04	.13		.56	-.34			.02	-.10	.24		.27	-.17	-.20	-.16				.20				-.02
REL	.03	.22	.16		.61	-.07			.16	.10	.23		.19	-.10	-.23	.08				.23				.18
SOC	.01	.26	.31		.73	-.26			.02	.05	.15		.08	.08	-.15	.06				.16		.06		.16
SP	-.03	.13	.49	-.14	.63	-.16	.12	-.31	-.13	.08	.20	.20	-.01	-.17	.67	-.19	.08	.61	.08	.16	.15	-.06	.00	.13
Avg	-.14	.00	.38	-.23	.54	.01	-.24	-.40	.24	-.13	.06	.01	-.08	-.30	.55	-.30	-.01	.51	-.05	.05	.02	-.18	-.16	-.13

in the classes that he or she takes. The value of a_i can be considered a rescaled version of student i's "true" GPA.

To account for the fact that students' achievements vary from class to class, suppose also that there is a distribution centered on each a_i that represents this variation. This distribution accounts for the fact that students don't always perform at the same level, and so don't always receive the same grade. This distribution also accounts for the fact that instructors seldom measure students' performances exactly.

Assuming that this distribution has the shape of a normal curve, the probability that any particular student gets a particular grade in a class can be represented graphically as in Figure 3. In that figure, the boxed regions under the curves represent the probabilities predicted by the model that the student would receive each of the grades depicted. For example, the top-most curve in Figure 3 represents the probability that a student would get a B in the given class, under the assumption that the student's distribution of "achievement" was centered around the bell-shaped curve depicted in the plot. More generally, in every class a student takes, and for every grade a student gets, the curves depicted in the top and bottom panels of Figure 3 are assumed to be centered on that student's average achievement, a_i.

The probability that a student receives any grade in a course is determined by the corresponding area under the curve and between that grade's thresholds. By shifting the center of the curve to the right, the probability that a student receives a higher grade is increased. Because the center of this curve represents the average level of a student's achievement, higher student achievement indices are assigned to students who receive higher grades. Heuristically, each student's achievement index is found by determining the location of these curves that is most consistent with the student's grades.

The vertical lines under the curves in Figure 3 represent grade thresholds. These thresholds depict the level of performance that an instructor implicitly used in assigning grades within his or her class. Because different instructors grade with varying degrees of grading

FIGURE 3

Positioning a student's achievement index so as to maximize the probability
predicted by the model that the student received grades of B, A−, and B+. In
this plot, the grade thresholds for the three classes the student took are the same
in both the top and bottom plots. The positioning of the curves on the top plot
(all centered on the same value of 3.3) results in higher areas (probabilities) for
the grades that the student received than does the positioning in the lower plot.
The achievement index of the student is determined by maximizing the product
of the areas assigned to each grade that the student received.

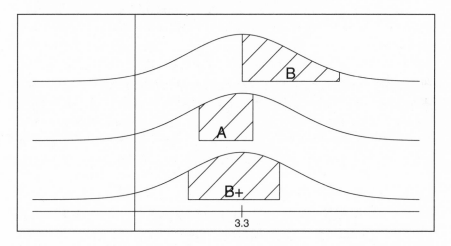

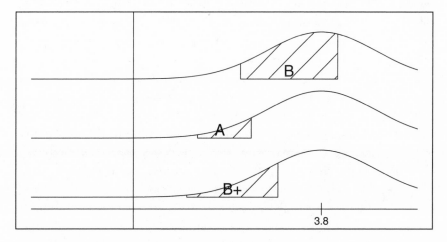

student achievement

stringency, these thresholds are estimated for every class and for every instructor.

The fundamental idea behind the achievement index is to estimate the mean academic performance for every student (a_i) and the grade thresholds for every class so as to maximize the probability of observing the grades that students actually receive. To see how this might be done, consider again the two plots depicted in Figure 3. The student whose grades are reflected in these plots received grades of B, A−, and A+ in the three courses that he or she took. Assuming that the grade thresholds in these classes have already been determined, then the top panel of plots represents the probability that the student received each of these grades when his or her achievement index was stipulated to be 3.3, while the bottom panel shows the corresponding probabilities when the student's achievement index was assumed to be 3.8. Since the product of the probabilities (boxed areas) in the top plot is greater, a value of 3.3 for the student's achievement index is favored to the value of 3.8 depicted in the lower panel. By moving the center of the student's curves to the right and left, the value of the student's achievement index that maximizes the product of these areas can be found. This is the value used to compute the student's adjusted GPA.

Now suppose that all student achievement indices have been estimated for a given set of grade thresholds. If the grade thresholds that were used to estimate these achievement indices aren't optimal, then better thresholds can be estimated in a fashion similar to that used above to estimate the achievement indices. This procedure is illustrated in Figure 4. In this case, the achievement indices for four students (i.e., the centers of the four curves) are held fixed, and the grade thresholds are shifted. Because the location of the thresholds in the top panel assigns a higher value to the product of the probabilities of the assigned grades than does the location of the thresholds in the bottom panel, that positioning of the thresholds is preferred. The optimal value of the thresholds, for given values of student achievement indices, is determined by positioning the grade thresholds so as to maximize the product of the areas representing the probabilities that each student received the grade they did.

FIGURE 4

Positioning of thresholds so as to maximize the predicted probability that the four students in this class received grades of A, A−, B+, and B+. In this plot, the achievement indices of the four students in the hypothetical class are the same in both the top and bottom panels. However, the product of the areas assigned to the grades that the students received in the top panel is greater than in the bottom panel, owing to the fact that the areas assigned to the first student's A and third student's B+ are roughly three times as great in the top panel as in the lower panel (and despite the fact that the probability assigned to the second student's A− roughly doubles). The optimal grade thresholds for this class are determined by maximizing this product for a given set of student achievements.

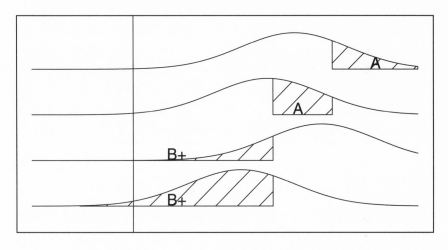

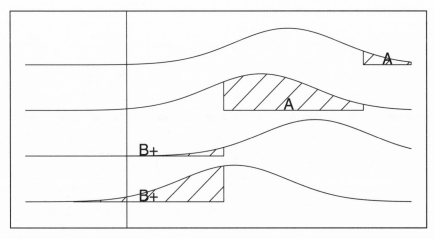

Joint estimates of the student achievement indices and grade thresholds can be found by iterating back and forth between estimating the achievement indices and grade thresholds until neither changes.

8 Conclusions

To MOST NON-ACADEMICS, THE EVIDENCE AND analyses presented in this book have probably seemed superfluous, serving only to prove assertions that are evident from common experience. Most people readily accept the ideas that different teachers employ different grading standards and that grading standards vary across disciplines. Few would argue that these differences do not cause students to take fewer courses from instructors who grade stringently, or that fewer students take science and math courses because of these differences. Most would agree that premed and science majors have to work harder to graduate with high GPAs than do humanities majors. And no one outside of academia really questions the supposition that teachers who grade leniently are likely to receive higher student evaluations.

Yet every one of these propositions has been challenged and dismissed by members of the academy. In part, the source of this conflict stems from a general distaste among faculty members for grading. Many professors, particularly those in the humanities, simply object to the requirement for them to grade students. As Wilson puts it, "the humanities have become hostile to hierarchy, and grading is inherently hierarchical" [Wil99, 42]. A fear that classroom rapport will be diminished probably also factors in. But regardless of one's philosophy toward grading, it must be acknowledged that faculty self-interest also plays a prominent role in this conflict.

Faculty have contrived an almost unending sequence of fables to avoid dealing with the unpleasantries associated with student grading. The teacher-effectiveness theory described in Chapter 3 provides an ideal example of such a myth. The teacher-effectiveness theory was fabricated by educational researchers to repudiate findings that student grades biased student evaluations of teaching, and has been used by professors ever since to justify the assignment of higher-than-average grades. In considering the validity of this theory, or, for that matter, any claim that grades have an effect contrary to what common sense would dictate, several facts concerning grades and grading should be recognized.

First, stringent grading demands more effort than lenient grading. In today's college environment, a professor cannot assign grades of C or D without tangible justification. Students become indignant when they receive such grades, and typically complain vigorously to administrators and (wealthy alumni) parents when they do. "Average" grades simply cannot be assigned without supporting evidence in the form of poor test marks, unsatisfactory homework scores, or low project grades. Generating this evidence requires time.

Consider the case of a professor teaching an introductory class containing, say, 100 students. One hundred students is a large class at Duke University, but at many public institutions it may be close to or even below the norm. Administering one quiz or exam to such a class requires several hours of exam preparation, and, for *each exam* submitted, between 15 minutes to one hour to grade. The time devoted to administering a single exam thus totals about 30 to 100 hours. In this respect, science and mathematics faculty have an advantage over their humanities and social science colleagues because exam items in their fields can be phrased so as to permit more efficient grading, often by a teaching assistant. Humanities faculty and, to a lesser extent, social science faculty are usually faced with the prospect of either grading lengthy essay problems or compromising the integrity of their exam by including relatively superficial multiple-choice or short-answer questions.

Class projects can be even worse. A colleague of mine regularly assigns end-of-term projects to her introductory statistics class. Requirements of these projects include 10-minute in-class presentations and written summaries. With over 100 students in her section, there is not sufficient time for students to present their work in the lecture sessions, and so for the last three to five weeks of the semester she attends each of her class's four weekly recitation sections to evaluate oral presentations. In addition, each project's written component must be graded. Clearly, such comprehensive student assessment requires an enormous commitment of time and energy.

The desire to establish rapport with students also causes instructors to grade leniently. Professors are intuitively aware of the consequences of grade-attribution theory. Students who receive lower than expected grades often blame their instructors for these grades, and it is hard to interact positively with students who are upset about their grades.

Another factor often used to explain differences in grading practices between the humanities and science departments is the inherently subjective nature of activity in the humanities versus the more objective nature of activity in the sciences. In mathematics, the argument goes, items on an exam usually have clear and unambiguously defined solutions. Partial credit, when given, can be based on the extent to which a student's solution differs from a correct solution. In contrast, problems in the humanities seldom have such well-defined solutions. Numerous answers, each having comparable merit, can be offered in response to the same question. The quality of student text can thus be viewed subjectively, depending as much on the perspective of the grader as the characteristics of the text itself. From this view, assigning grades becomes largely an arbitrary exercise.

This explanation for the observed differences in grading practices between the humanities and sciences fails, however, when it is used to justify high grades for all. The fact that an instructor is unable to distinguish between the quality of work from different students does not mean that all students should receive equally high grades. It only means that all students should receive equal grades. That is, if an instructor insists that all students were able "to display with greater diversity a legitimate and appropriate grasp of a widened context" [Bil95, 443], then, by definition, all students performed at the *average* level and so should all be given *average* grades.

Regardless of the source of these and other academic myths that contribute to the misunderstanding of the effects of grading and assessment in postsecondary education, the time has come to dispose of these fables. Research findings presented in this book, along with seventy years of previous educational research, permit the following conclusions to be drawn:

1. Differences in grading practices between instructors cause biases in student evaluations of teaching.
2. Student evaluations of teaching are not reliable indicators of teaching effectiveness and account for only a small proportion of the variance in student learning from student to student and course to course.
3. High grade distributions cannot be associated with higher levels of student achievement.
4. Differences in grading practices have a substantial impact on student enrollments, and cause fewer students to enroll in those fields that grade more stringently.
5. Grading practices differ systematically between disciplines and instructors, and these disparities cause serious inequities in student assessment.

The case that grading practices differ systematically between disciplines was established in Chapter 7 using the historical comparisons of departmental grading practices initiated by Goldman and Widawsi [GW76], continued by Strenta and Elliot [SE87, ES88], and extended here. The basis for each of these analyses was the examination of grades received by the same student in courses taught in different departments. By basing their analyses on pairwise comparisons, Goldman, Widawski, Strenta, and Elliot effectively eliminated any possibility that observed differences in grades assigned by different departments could be attributed to differences in the general abilities of students. The analysis presented in Chapter 7, in which pairwise differences were computed separately for students in different majors, went further by eliminating the possibility that these differences could be ascribed even to special abilities. That grading practices differ systematically between disciplines is beyond dispute. Humanities faculty tend to grade most leniently; social sciences faculty, with the exception of economists, are approximately grade neutral; and economics, mathematics, and natural sciences faculty tend to grade most stringently.

Effects of differential grading practices on student enrollment decisions were investigated in Chapter 6. Surreptitious observations made of students as they navigated the DUET website to acquire information about mean grades of classes taught in past semesters were combined with DUET survey data and subsequent enrollment data to estimate the effect of grading policy on the probability that a student enrolled in a course. In simple terms, the primary finding of this chapter was that students at Duke University were approximately twice as likely to enroll in an elective course expected to be graded at an A− average as they were to enroll in a course that they expected to be graded at a B average. This result applies both to student selection of courses within similar academic disciplines and to student selection of instructors for a given course. From an educational policy perspective, the implications of this finding are startling: As a consequence only of differences in grading practices between academic fields, American undergraduates take, on average, about 50% fewer elective courses in the natural sciences and mathematics than they would if grading practices across disciplines were more equitable. The implications of this fact for the nation's economy are unclear. What is clear, however, is the fact that there would be a substantial shift of educational resources from humanities departments to natural sciences and mathematical sciences departments, and a substantial increase in the preparation of college graduates to participate in an increasingly technological society, if only grades were assigned more equitably across academic disciplines.

One of the longest-running debates in educational research involves the question of whether and to what extent the grading practices of an instructor affect students' evaluations of teaching. Although most researchers now concede that higher grades do translate into higher course evaluations, many researchers continue to insist that the magnitude of this relationship is small. This insistence often stems from a researcher's desire to demonstrate that their survey "instrument" accurately reflects the teaching dimensions that it is intended to measure, and that biases resulting from external factors—like student grades—are small.

By presenting the first large-scale study in which the effects of grades on student evaluations of teaching were examined by contrasting individual student responses elicited both before and after receipt of final course grades, the analysis in Chapter 4 sheds new light on this debate. The conclusion of this analysis is that students' grades in a course—whether they be the grades they expect to receive or actually do receive—have a significant impact on their responses to teacher-course evaluation forms. This conclusion applies to all teaching-related items included on the DUET survey and corroborates the findings of experiments conducted in the late 1970s and early 1980s in which student grades were artificially manipulated. It is also consistent with a preponderance of observational studies that have examined this phenomenon. Student evaluations of teaching, if not corrected for grading practices (and a variety of other factors like class size, course level, and GPAs of students), represent a biased measure of the quantity they attempt to measure.

Of course, the group most directly affected by grading disparities are students. The example transcripts presented in Chapter 7 clearly illustrate the seriousness of this problem and demonstrate that for students in the middle half of an academic class, grading practices of instructors are as important in determining class rank as their performance in class. Indeed, for the Duke class whose grades were considered in Chapter 7, the standard deviation of the difference between conventional GPA and achievement-index GPA for the middle half of students was 63% as large as the standard deviation of the GPA itself. Knowing the grading practices of the instructors from whom students took courses is as important as knowing the grades they got.

REFORM

Disparities in student grading have led to a general degradation of America's postsecondary educational system. Inconsistent grading standards result in unjust evaluation of students and faculty, and discourage students from taking those courses that would be of greatest benefit to them. It is somewhat paradoxical that given the vast resources devoted each year to improving education within our colleges, so basic a problem has remained unrepaired for so long. Why, then, has this problem persisted for so long, and how can the system be reformed?

At most universities, faculty senates are responsible for over-seeing campus policies that affect academics. In the normal course of events, these legislative bodies would be responsible for imple-menting reforms to university grading policies. Unfortunately, many faculty councils are populated by the very individuals who benefit most from inequitable grading policies. Or they represent departments that do. Appropriate strategies for reform thus depend on the integrity of the individuals who compose these bodies. Pro-vided that members of these governing faculty bodies are willing to enact changes to reverse grading inequities, even if these changes are unpopular among their constituencies, then there is hope that reform can be coordinated in a cooperative way among faculty mem-bers at the institutional level. This process can also be facilitated by administrative officers responsible for the formation and oversight of committees involved in faculty reviews and in establishing academic standards.

Assuming adequate faculty and administrative support, any number of strategies might be adopted to enact reform. Rosovsky and Hartley [RH02] outlined several of these in their recent report on grade inflation. Since these policies are also appropriate for correcting the more basic problem of disparities in grading practices, I repeat them here.

1. Encourage institutional dialogue. Few colleges or universi-ties delineate grading policies to incoming faculty members. Organized faculty discussions of grading practices are equally rare. Because discussion of grading practices is a prerequisite to reform, dialogue is critical.

   Provosts and deans can play a pivotal role in establishing this dialogue by communicating their concerns to department chairs and other faculty leaders, who in turn can relay these concerns and initiate further discussion within their academic units. Alternatively, faculty committees and legislative bod-ies can initiate discussion through the organization of forums and study groups. Student input should also be solicited.

2. Provide instructors with more information about their uni-versity's grading practices. Many professors do not know

how their grade distributions compare to their colleagues', or how their departmental standards compare to others'.

This problem can be partially eliminated by providing professors with summaries of the grade distributions assigned by other faculty and departments. Duke University regularly does this, although the success of this practice there has been limited. More sophisticated approaches to alleviating this problem might involve providing instructors with individualized assessments of their grading practices. For example, standardized grade intervals similar to those displayed in Tables 7 and 8, or the adjusted grades displayed in Tables 5 and 6 of Chapter 7, might be appended to departmental and institutional grade summaries. This would assist instructors in quantitatively assessing their grading practices in relation to institutional norms.

3. Constrain course grade distributions. This practice is fairly common in graduate and professional schools, and both the School of Law and Fuqua School of Business at Duke University use this method to maintain grading standards. For example, as late as 1995 Duke's Law School imposed a median grade constraint on grades assigned in all classes containing more than 30 students.

Such policies can, however, create their own inequities. Within the Duke law school, the use of median grade constraints accentuated differences between the grading practices in classes having more than 30 students and those having fewer than 30 students. Other problems evolved as a consequence of the distinction between the statistical definition of the median and mean. Apparently, professors in the law school knew the difference between these two statistics, and many adopted a grading strategy to fully exploit it; several professors regularly assigned approximately 51% of their grades exactly at or just below the upper limit on the median grade, and assigned their remaining grades at arbitrarily high levels.

The problems associated with the simplistic grade constraints employed by the Duke law school illustrate some of the difficulties inherent in establishing coherent grading policies. Care must be taken to ensure that constraints transition smoothly from large classes to small and that the mathematical properties of the constraint be well understood. Constraints should be specified to afford instructors as much flexibility as possible without, of course, compromising the purpose of the constraint.

A more successful strategy for imposing constraints on grade distributions might be formulated by requiring mean course grades to fall within, say, δ grade units of a mean value, say μ. Here, μ represents a weighted average of a target grade level and an estimate of the "true" GPA of students in a class. For example, suppose that university faculty and administrators agree upon a target mean grade of B, or a 3.0. Then the mean grade constraint for individual classes might be determined using a formula similar to

$$\mu = w \times 3.0 + (1 - w) * g, \qquad (8.1)$$

where w is a weight constrained to lie in the interval $(0,1)$. The value g is either the mean GPA of students entering the course or an adjusted mean GPA. For first-year courses or other large introductory classes, w might be set to 1, making the target grade for a class equal to 3.0. In advanced seminar courses composed of students whose GPAs or adjusted GPAs were well established, w might be assigned a value closer to 0, resulting in a mean course grade constraint that was close to the incoming mean GPA of its students.

The value of δ determines the flexibility given to an instructor in adhering to the constraint. For large introductory classes in which there is little reason to believe that the abilities of students differ greatly from the general student population, δ might be fixed at some minimum value, say 0.05. For a target constraint of 3.0, any mean course grade between 2.95 and 3.05 would then be accepted. In smaller classes, δ would be increased to account for greater variation in the mean achievement levels of students within the class. Larger values of δ in smaller classes are also required to account for discreteness in the averaging process and to avoid situations in which the grade of say, a single student, is completely determined by the grading constraint.

A potential difficulty with this form of constraint is that the contribution, g, to the course mean grade may itself represent a distorted measure of student achievement levels. One source of contamination in g arises from the necessary flexibility built into the scheme, namely, that instructors in small classes are permitted to assign higher than average grades. As a result, students who take an unusually high proportion of seminar courses or independent study classes might accrue artificially high GPAs. Adjusting course grade constraints based on their GPAs might only serve to exacerbate the problem.

Another source of distortion in g arises from nonrandom assignment of students to first-year classes. If, for example, the second semester GPAs of engineering students were determined primarily from grades received in large, first-semester, introductory engineering courses populated by above-average engineering students, then a lower than expected GPA constraint could propagate through subsequent courses taken by these students.

Most of these difficulties can be avoided through the use of adjusted GPAs to determine g. For the achievement-index adjusted GPA, distortions attributable to small classes would be minimized, and, in the case of independent study courses, completely eliminated.

4. Include information about course grading practices on student transcripts. As Rosovsky and Hartley [RH02, 15] point out, such policies have been adopted by Columbia, Dartmouth, Indiana, and Eastern Kentucky, although, as mentioned in Chapter 1, the success of these policies has been questionable.

In my opinion, carefully designed constraints on mean course grades provide the most comprehensive solution to problems associated with disparities in grading practices. Such constraints clearly eliminate incentives for students to take courses from instructors who grade leniently. In addition, they naturally eliminate biases to student evaluations of teaching caused by differential grading practices. Finally, by tying grade constraints to mean performance levels of students registered for a class, incentives for students to take courses populated by below-average students, or conversely, disincentives for students to take courses populated by above-average students, are avoided.

Unfortunately, the notion of constraining mean grades in classes has not proven palatable to the faculties of most major universities. Likewise, initiating faculty discussions of grading, providing mean grade information on transcripts, and supplying faculty with information about the grading practices of their colleagues has not yet led to meaningful reform. Given the relative failure of these strategies, I would propose several additional possibilities. If imposing constraints on class mean grades is considered the most radical of the suggested strategies listed above, and faculty discussions on grading the least radical, then these proposals might be considered centrist.

They are:

5. Allow students to optionally report adjusted GPAs on their transcripts. Because both of the GPA adjustment methods described in Chapter 7 result in demonstrably more accurate summaries of student performance, there is no reason why students should not have the option of reporting these summaries of their academic performance in place of traditional GPA.

    In the short term, this policy would cause a spike in inflationary grading trends because students would choose to report the more favorable summary of their performance. In the longer term, however, it might lead some students to pressure leniently grading faculty to be more discerning in their grade assignments. Savvy recruiters might also catch on, and begin requesting adjusted GPAs. This, too, would result in downward pressure on grades.

6. Use adjusted grades and GPAs to establish honors distinctions. An important advantage of grade adjustment methods is that they do not require explicit changes in the grading practices of individual faculty. The use of adjusted GPAs would, however, require that faculty differentiate between the performance of those students who they feit were most deserving of honors and those who were not.

   To understand why this policy would force faculty to reserve top grades for those students most deserving of honors, recall that under the achievement index method for adjusting grades, instructors who assign A's to all students in a class do not affect the adjusted GPA of any of their students. More generally, instructors who grade leniently have less impact on the adjusted GPAs of their top students than do instructors who grade stringently. Additive adjustments to grades have a similar effect. Thus, faculty who wish to affect the honors process would have to more carefully dispense with their highest marks.

   A related benefit of basing undergraduate honors on adjusted GPAs was recently suggested to me at a faculty forum on grading held at Smith College. By encouraging faculty to reserve their highest marks for top students, this practice would help re-establish the A as a grade reserved for truly outstanding performance.

7. Selectively exclude student evaluations of teaching from instructor summaries. Student evaluations of teaching play an important role in faculty promotion, tenure, and salary reviews and as discussed in Chapters 3 and 4, student evaluations of teaching are affected by the grades faculty assign to students. Without adjustment, these forms provide a strong incentive for the assignment of high grades, and a disincentive for the assignment of low grades. To eliminate the disincentive for assigning low grades, some proportion of an instructor's student evaluations might be ignored, with this proportion depending on the grade distribution employed by

the instructor. For example, in a class in which 10% of the assigned grades were C's or lower, the lowest 10% of student evaluation scores might be ignored. Similarly, the lowest 20% of course evaluations would be excluded from summaries of instructors who assigned 20% C grades or lower.

Conversely, some proportion of the best student evalua-tions of teaching might be excluded from student evaluations of teaching for courses in which the number of A's exceeded some target threshold. For example, if faculty and staff agreed to a target of, say, 20% A's in every class, and an instructor assigned 30% A's, then the best 10% of his course evaluations would be excluded from administrative reviews. Although such exclusions clearly overcorrect for the effect of grades on student evaluations of teaching, they would undoubtedly prove useful in altering grade distributions.

The final three recommendations have the advantage of being minimally invasive; they can be enacted without requiring coopera-tion from every member of a university faculty. Yet each represents a concrete first step down the road to grading reform.

A crisis exists. Current assessment practices are flawed, and both students and faculty know it. Unregulated grading practices change student enrollment patterns and penalize students who pursue demanding curricula. They permit students to manipulate their GPAs through the judicious choice of their classes rather than through the moderation of their performance in those classes. Disparities in grading also affect the way students complete end-of-course evaluation forms, and so result in inequitable faculty assess-ments. As a consequence, academic standards are diminished.

To right the boat, two things must happen: More principled stu-dent grading practices must be adopted, and faculty assessment must be more closely linked to student achievement.

It is my hope that this book facilitates reform in both areas.

Bibliography

[Abr85]   P.C. Abrami. Dimensions of effective college teaching. *Review of Higher Education*, 8:211–228, 1985.

[Ad90]   P.C. Abrami and S. d'Apollonia. The dimensionality of student ratings and their use in personnel decisions. In M. Theall and J. Franklin, editors, *Student Ratings of Instruction: Issues for Improving Practice.*, volume 43 of *New Directions for Teaching and Learning*, pages 97–111. Jossey-Bass, San Francisco, 1990.

[Ad91]   P.C. Abrami and S. d'Apollonia. Multidimensional students' evaluations of teaching effectiveness—generaliizability of "$n = 1$" research: Comment on Marsh (1991). *Journal of Educational Psychology*, 83:411–415, 1991.

[AdR97]   P.C. Abrami, S. d'Apollonia, and S. Rosenfield. The dimensionality of student ratings of instruction: What we know and what we do not know. *Higher Education: Handbook of Theory and Research*, 11:213–264, 1997.

[ALP82]   P.C. Abrami, L. Leventhal, and R.P. Perry. Educational seduction. *Review of Educational Research*, 52:446–464, 1982.

[Ani53]   A.M. Anikeef. Factors affecting student evaluation of faculty members. *Journal of Applied Psychology*, 37:458–460, 1953.

[AR93]   N. Ambady and R. Rosenthal. Half a minute: Predicting teacher evaluations from thin slices of nonverbal behavior and physical attractiveness. *Journal of Personality and Social Psychology*, 64:431–441, 1993.

[AS73]   L.M. Aleamoni and R.E. Spencer. The Illinois course evaluation questionnaire: a description of its development and a report of some of its results. *Educational and Psychological Measurement*, 33:669–684, 1973.

[Ben53]   A.W. Bendig. Relation of level of course achievement of students, instructors, and course ratings in introductory psychology. *Journal of Educational Psychology*, 13:487–488, 1953.

[Bil95]   D. Bilimoria. Modernism, postmodernism, and contemporary grading practices. *Journal of Management Education*, 19:440–458, 1995.

[Bla74]   T. Blass. Measurement of objectivity–subjectivity: effects of tolerance for imbalance and grades on evaluations of teachers. *Psychological Reports*, 34:1199–1213, 1974.

[Blu36]   M.L. Blum. An investigation of the relation existing between students' grades and their ratings of their instructors ability to teach. *Journal of Educational Psychology*, 27:217–221, 1936.

[Blu91]   A. Blunt. The effects of anonymity and manipulated grades on student ratings of instructors. *Community College Review*, 18:48–54, 1991.

[BM72]   R.B. Bausell and J. Magoon. Expected grade in a course, grade point average, and student ratings of the courses and the instructor. *Educational and Psychological Measurement*, 32:1013–1023, 1972.

[Bro76]   D.L. Brown. Faculty ratings and student grades: a university-wide multiple regression analysis. *Journal of Educational Psychology*, 68:573–578, 1976.

[BS76]   R.P. Barnoski and A.L Sockloff. A validation study of the faculty and course evaluation (face) instrument. *Educational and Psychological Measurement*, 36:391–400, 1976.

[BSB77]   D.C. Brandenburg, J.A. Slinde, and E.E. Batiste. Student ratings of instruction: validity and normative interpretations. *Research in Higher Education*, 7:67–78, 1977.

[Caf69]   B. Caffrey. Lack of bias in student evaluations of teachers. *Proceedings of the 77th Annual Convention of the American Psychological Association*, 4:641–642, 1969.

[Cen75]   J.A. Centra. Colleagues as raters of classroom instruction. *Journal of Higher Education*, 46:327–337, 1975.

[Cen77]   J.A. Centra. Student ratings of instruction and their relationship to student learning. *American Educational Research Journal*, 14:17–24, 1977.

[Cen79]   J.A. Centra. *Determining Faculty Effectiveness*. Jossey-Bass, 1979.

[CGM71]   F. Costin, W.T. Greenough, and R.J. Menges. Student ratings of college teaching: reliability, validity, and usefulness. *Review of Educational Research*, 41(5):511–535, 1971.

[Cha83]   T.I. Chacko. Student ratings of instruction: A function of grading standards. *Educational Research Quarterly*, 8(2):19–25, 1983.

[CL73]   J.A. Centra and R.L. Linn. Student points of view in ratings of college instruction. Technical Report RB-73-60, Educational Testing Service, 1973.

[CLW96]   J. Caulkin, P. Larkey, and J. Wei. Adjusting GPA to reflect course difficulty. Technical report, Heinz School of Public Policy and Management, Carnegie Mellon University, 1996.

[CM81]   J. Coleman and W.J. McKeachie. Effects of instructor/course evaluations on student course selection. *Journal of Educational Psychology*, 73:224–226, 1981.

[CN75]   K.S. Crittenden and J.L. Norr. Some remarks on student ratings: The validity problem. *American Educational Research Journal*, 12:429–433, 1975.

[Coh80]   P.A. Cohen. Effectiveness of student-rating feedback for improving college instruction. *Research in Higher Education*, 13:321–341, 1980.

[Coh81]   P.A. Cohen. Student ratings of instruction and student achievement. *Review of Educational Research*, 51:281–309, 1981.

[Coh87]   P.A. Cohen. A critical analysis and reanalysis of the multisection validity analysis. In *1987 Annual Meeting of the American Educational Research Association (ERIC Document Reproduction Service No. ED 283 876)*, 1987.

[Col93]   W. Cole. The perils of grade inflation. *Chronicle of Higher Education*, pages B1–B2, January 6, 1993.

[Com99]   A. Comarow. Grades are up, standards are down. *U.S. News and World Report America's Best Colleges*, 1999.

[Dom71]   G. Domino. Interactive effects of achievement orientation and teaching style on academic achievement. *Journal of Educational Psychology*, 62:427–431, 1971.

[Doy75]   K.O. Doyle. *Student Evaluation of Instruction*. Heath, Lexington, MA, 1975.

[EDP76]   G.T. Endo and G. Della-Piana. A validation study of course evaluation ratings. *Improving College and University Teaching*, 24:84–86, 1976.

[ES88]   R. Elliot and A.C. Strenta. Effects of improving the reliability of the GPA on prediction generally and on comparative predictions for gender and race particularly. *Journal of Educational Measurement*, 25:333–347, 1988.

[Fel76]   K.A. Feldman. Grades and college students' evaluations of their courses and teachers. *Research in Higher Education*, 4:69–111, 1976.

[Fel88]   K.A. Feldman. Effective college teaching from the students' and faculty's view: Matched or mismatched priorities? *Research in Higher Education*, 28:291–344, 1988.

[FLB75]   P.W. Frey, D.W. Leonard, and W.W. Beatty. Student ratings of instruction: Validation research. *American Educational Research Journal*, 12:435–444, 1975.

[Fre73]   P. Frey. Student ratings of teaching: validity of several rating factors. *Science*, 182:83–85, 1973.

[Fre76]   P. Frey. Validity of student instructional ratings as a function of their timing. *Journal of Higher Education*, 47:327–338, 1976.

[GB90]   R.J. Gigliotti and F.S. Buchtel. Attributional bias and course evaluations. *Journal of Educational Psychology*, 82:341–351, 1990.

[GC62]   C.M. Garverick and H.D. Carter. Instructor ratings and expected grades. *California Journal of Educational Research*, 13:218–221, 1962.

[Ges73]   P.K. Gessner. Evaluation of instruction. *Science*, 180:566–569, 1973.

[GG97]   A.G. Greenwald and G.M Gillmore. Grading leniency is a removable constraint. *American Psychologist*, 52:1209–1217, 1997.

[GHRPS90]   A.E. Gelfand, S.E. Hills, A. Racine-Poon, and A.F.M. Smith. Illustration of Bayesian inference in normal data models using Gibbs sampling. *Journal of the American Statistical Association*, 85:398–409, 1990.

[GP73]   K.L. Granzin and J.J. Painter. A new explanation for students' course evaluation tendencies. *American Educational Research Journal*, 8:435–445, 1973.

[Gre97]   A.G. Greenwald. Validity concerns and usefulness of student ratings of instruction. *American Psychologist*, 52:1182–1186, November 1997.

[GSHF74]   R.D. Goldman, D.E. Schmidt, B.N. Hewitt, and R. Fisher. Grading practices in different major fields. *American Education Research Journal*, 11:343–357, 1974.

[GW76]   R.D. Goldman and M.H. Widawski. A within-subjects technique for comparing college grading standards: implications in the validity of the evaluation of college achievement. *Educational and Psychological Measurement*, 36:381–390, 1976.

[Hel47]   H. Helson. Adaptation-level as frame of reference for prediction of psychophysical data. *American Journal of Psychology*, 60:1–29, 1947.

[Hel48]　H. Helson. Adaptation-level as a basis for quantitative theory of frame of reference. *Psychological Review*, pages 297–313, 1948.

[HM80]　G.S. Howard and S.E. Maxwell. Correlation between student satisfaction and grades: a case of mistaken causation? *Journal of Educational Psychology*, 72:810–820, 1980.

[Hoc76]　J.M. Hocking. College students' evaluations of faculty are directly related to course interest and grade expectation. *College Student Journal*, pages 312–316, 1976.

[Hol71]　D.S. Holmes. The relationship between expected grades and students' evaluations of their instructor. *Educational and Psychological Measurement*, 31:951–957, 1971.

[Hol72]　D.S. Holmes. Effects of grades and disconfirmed grade expectancies on students' evaluation of their instructor. *Journal of Educational Psychology*, 63:130–133, 1972.

[Hud51]　E. Hudelson. The validity of student rating of instructors. *School and Society*, 73:265–266, 1951.

[Joh97]　V.E. Johnson. An alternative to traditional GPA for evaluating student performance. *Statistical Science*, 12:257–278, 1997.

[Joh99]　V.E. Johnson and J. Albert. *Ordinal Data Modeling*. Springer-Verlag, New York, NY, 1999.

[Joh02]　V.E. Johnson. Teacher course evaluations and student grades: An academic tango. *Chance*, 15(3):9–16, 2002.

[Kap74]　D.E. Kapel. Assessment of a conceptually based instructor evaluation form. *Research in Higher Education*, 2:1–24, 1974.

[Kel72]　A. Kelley. Uses and abuses of course evaluations as measures of educational output. *Journal of Economic Education*, pages 13–18, 1972.

[Ken75]　W.R. Kennedy. Grades expected and grades received—their relationship to students' evaluations of faculty performance. *Journal of Educational Psychology*, 67:109–115, 1975.

[KK76]   R. Kovaks and D.E. Kapel. Personality correlates of faculty and course evaluations. *Research in Higher Education*, 5:335–344, 1976.

[KM75]   J.A. Kulik and W.J. McKeachie. The evaluation of teachers in higher education. *Review of Research in Education*, 3:210–240, 1975.

[KM95]   J. Koon and H. Murray. Using multiple outcomes to validate student ratings of overall teacher effectiveness. *Journal of Higher Education*, 66:61–81, 1995.

[Koo68]   E.W. Kooker. The relationship of known college grades to course ratings on student selected items. *Journal of Psychology*, 69:209–215, 1968.

[KS99]   A.C. Krautmann and W. Sander. Grades and student evaluations of teachers. *Economics of Education Review*, 18:59–63, 1999.

[Lan99]   R.E. Landrum. Student expectations of grade inflation. *Journal of Research and Development in Education*, 32:124–128, 1999.

[LC92]   P. Larkey and J. Caulkin. Incentives to fail. Technical Report 92-51, Heinz School of Public Policy and Management, Carnegie Mellon University, 1992.

[Man01]   H. Mansfield. Grade inflation: It's time to face the facts. *Chronicle of Higher Education*, page B24, April 6, 2001.

[Mar80]   J.U. Marsh, H.W. Overall. Validity of students' evaluations of teaching effectiveness: Cognitive and affective criteria. *Journal of Educational Psychology*, 72:468–475, 1980.

[Mar82]   H.W. Marsh. SEEQ: A reliable, valid, and useful instrument for collecting students' evaluations of university teaching. *British Journal of Educational Psychology*, 52:77–95, 1982.

[Mar83]   H.W. Marsh. Multidimensional ratings of teaching effectiveness by students from different academic settings and their relation to student/course/instructor characteristics. *Journal of Educational Psychology*, 75:150–166, 1983.

[Mar84]   H.W. Marsh. Students' evaluations of university teaching: dimen-
           sionality, reliability, validity, potential biases, and utility. *Journal of
           Educational Psychology*, 76:707–754, 1984.

[MBS56]   J.E. Morsh, G.G. Burgess, and P.N. Smith. Student achievement
           as a measure of instructor effectiveness. *Journal of Educational
           Psychology*, 47:79–88, 1956.

[McC80]   P. McCullagh. Regression models for ordinal data. *Journal of the
           Royal Statistical Society, Series B*, 42:109–142, 1980.

[McK79]   W.J. McKeachie. Student ratings of faculty. *Academe*, 65:384–397,
           1979.

[McK97]   W.J. McKeachie. Student ratings: Validity of use. *American Psy-
           chologist*, 52:1218–1225, November 1997.

[MD92]   H.W. Marsh and M.J. Dunkin. Students' evaluations of univer-
           sity teaching: A multidimensional perspective. *Higher Education:
           Handbook of Theory and Research*, 8:142–233, 1992.

[MF79]   R.S. Meier and J.F. Feldhusen. Another look at Dr. Fox: Effect of
           stated purpose for evaluation, lecturer expressiveness, and density
           of lecture content on student ratings. *Journal of Educational Psy-
           chology*, 71:339–345, 1979.

[Mil72]   R.I. Miller. *Evaluating Faculty Performance*. Jossey-Bass, San
           Francisco, 1972.

[Mir73]   R. Mirus. Some implications of student evaluation of teachers.
           Journal of Economic Education, 5:35–37, 1973.

[MR93]   H.W. Marsh and L.A. Roche. The use of student evaluations
           and an individually structured intervention to enhance univer-
           sity teaching effectiveness. *American Educational Research Journal*,
           30:217–251, 1993.

[Nea96]   I. Neath. How to improve your teaching evaluations without
           improving your teaching. *Psychological Reports*, 78:1363–1372,
           1996.

[NL84]   J.P. Nelson and K.A. Lynch. Grade inflation, real income, simul-
           taneity, and teaching evaluations. *Journal of Economic Education*,
           21–37, 1984.

[NWD73]   D.H. Naftulin, J.E. Ware, and F.A. Donnelly. The Doctor Fox
            lecture: A paradigm of educational seduction. *Journal of Medical
            Education*, 48:630–635, 1973.

[Owi85]   I. Owie. Incongruence between expected and obtained grades and
            students' rating of the instructor. *Journal of Instructional Psychol-
            ogy*, 12:196–199, 1985.

[Pas79]   P.J. Pascale. Knowledge of final grade and effect on student eval-
            uation of instruction. *Educational Research Quarterly*, 4(2):52–57,
            1979.

[PC80]   C. Peterson and S. Cooper. Teacher evaluation by graded and
            ungraded students. *Journal of Educational Psychology*, 72:682–685,
            1980.

[Per98]   N. Perrin. How students at Dartmouth came to deserve better
            grades. *Chronicle of Higher Education*, page A68, October 9, 1998.

[Poh75]   J.T. Pohlmann. A multivariate analysis of selected class charac-
            teristic and student ratings of instruction. *Multivariate Behavioral
            Research*, 10:81–92, 1975.

[PP76]   M. Pratt and T.A.E.C. Pratt. A study of student–teacher grad-
            ing interaction process. *Improving College and University Teaching*,
            24:73–81, 1976.

[Rao73]   C.R. Rao. *Linear Statistical Inference and Its Applications, 2nd Edi-
            tion*. John Wiley & Sons, New York, 1973.

[Ray68]   N.F. Rayder. College student ratings of instructors. *Journal of
            Experimental Education*, 37:76–81, 1968.

[RB53]   H.E. Russell and A.W. Bendig. An investigation of the relation of
            student ratings of psychology instructors to their course achieve-
            ment when academic aptitude is controlled. *Educational and Psy-
            chological Measurement*, 13:626–635, 1953.

[RCF73]   B. Rosenshine, A. Cohen, and N. Furst. Correlates of student
            preference ratings. *Journal of College Student Personnel*, 14:269–
            272, 1973.

[Rem28]   H.H. Remmers. The relationship between students' marks and student attitude toward instructors. *School and Society*, 759–760, 1928.

[Rem30]   H.H. Remmers. To what extent do grades influence student ratings of instructors? *Journal of Educational Research*, 21:314–316, 1930.

[RH02]   H. Rosovsky and M. Hartley. Evaluation and the academy: Are we doing the right thing? American Academy of Arts and Sciences, 136 Irving Street, Cambridge, MA 02138-1996, 2002.

[RM70]   J. Rubenstein and H. Mitchell. Feeling free, student involvement, and appreciation. *Proceedings of the 78th Annual Convention of the American Psychological Association*, 5:623–624, 1970.

[RME49]   H.H. Remmers, F.D. Martin, and D.N. Elliot. Are students' ratings of instructors related to their grades? *Purdue University Studies in Higher Education*, 66:17–26, 1949.

[Rod75]   M. Rodin. Student evaluation. *Science*, 187:555–557, 1975.

[RR72]   M. Rodin and B. Rodin. Student evaluations of teachers. *Science*, 177:1164–1166, 1972.

[RRL50]   J.W. Riley, B.F. Ryan, and M. Lifschitz. *The Student Looks at His Teacher: An Inquiry into the Implications of Student Ratings at the College Level*. Rutgers University Press, New Brunswick, N.J., 1950.

[Sav45]   B.M. Savage. Undergraduate ratings of courses in Harvard College. *Harvard Educational Review*, 15:168–172, 1945.

[SC73]   A.J. Schuh and M.A. Crivelli. Animadversion error in student evaluations of faculty teaching effectiveness. *Journal of Applied Psychology*, 58:259–260, 1973.

[Sch75]   D.P. Schwab. Course and student characteristic correlates of the course evaluation instrument. *Journal of Applied Psychology*, 60(6):742–747, 1975.

[SD65]   R.E. Spencer and W. Dick. Course evaluation questionnaire: manual of interpretation. Technical Report 200, Measurement and Research Division, Office of Instructional Resources, University of Illinois, Urbana, 1965.

[SE87]   A.C. Strenta and R. Elliot. Differential grading standards revisited. *Journal of Educational Measurement*, 24:281–291, 1987.

[Sei83]   D.A. Seiver. Evaluations and grades: a simultaneous framework. *Journal of Economic Education*, 32–38, 1983.

[SF79]   S.A. Stumpf and R.D. Freedman. Expected grade covariation with student ratings of instruction: individual versus class effects. *Journal of Educational Psychology*, 71:293–302, 1979.

[SGD83]   V. Scheurich, B. Graham, and M. Drolette. Expected grades versus specific evaluations of the teacher as predictors of students' overall evaluation of the teacher. *Research in Higher Education*, 19:159–173, 1983.

[Sha90]   E.G. Shapiro. Effect of instructor and class characteristics on students' class evaluations. *Research in Higher Education*, 31:135–148, 1990.

[SM66]   C.T. Stewart and L.F. Malpass. Estimates of achievement and ratings of instructors. *Journal of Educational Research*, 59:347–350, 1966.

[Spe68]   R.E. Spencer. The Illinois course evaluation questionairre: Manual of interpretation. Technical Report 270, Office of Instructional Resources, University of Illinois, Champaign, IL., 1968.

[SS74]   A.M. Sullivan and G.R. Skanes. Validity of student evaluation of teaching and the characteristics of successful instructors. *Journal of Educational Psychology*, 66:584–590, 1974.

[Sta34]   J.A. Starrak. Student rating of instruction. *Journal of Higher Education*, 5:88–90, 1934.

[SWL91]   R. Sabot and J. Wakeman-Linn. Grade inflation and course choice. *Journal of Economic Perspectives*, 5:159–171, 1991.

[TF70]   D.J. Treffinger and J.F. Feldhusen. Predicting students' ratings of instruction. In *Proceedings of the 78th Annual Convention of the American Psychological Association*, 621–622, 1970.

[VF60]   V.W. Voeks and G.M. French. Are student ratings of teachers affected by grades? *Journal of Higher Education*, 31:330–334, 1960.

[VS79]   R. Vasta and R.F. Sarmiento. Liberal grading improves evaluations but not performance. *Journal of Educational Psychology*, 71:207–211, 1979.

[War73]   W.G. Warrington. Student evaluation of instruction at Michigan State University. In A.L. Sockloff, editor, *Proceedings: The First Invitational Conference on Faculty Effectiveness as Evaluated by Students*, 1973.

[WC97]   W.M. Williams and S.J. Ceci. How'm I doing? *Change*, September:13–23, 1997.

[Wea60]   C.H. Weaver. Instructor rating by college students. *Journal of Educational Psychology*, 51:21–25, 1960.

[Wil99]   B.P. Wilson. The phenomenon of grade inflation in higher education. *National Forum*, 79:38–49, 1999.

[Win77]   J.L. Winsor. A's, B's, but not C's?: A comment. *Contemporary Education*, 48(2):82–84, 1977.

[WW75]   J.E. Ware and R.G. Williams. The Dr. Fox effect: A study of lecturer effectiveness and ratings of instruction. *Journal of Medical Education*, 50:149–156, 1975.

[WW76]   R.G. Williams and J.E. Ware. Validity of student ratings of instruction under different incentive conditions. *Journal of Educational Psychology*, 68:48–56, 1976.

[WW77]   R.G. Williams and J.E. Ware. An extended visit with Dr. Fox: Validity of student satisfaction with instruction ratings after repeated exposures to a lecturer. *American Educational Research Journal*, 14:449–457, 1977.

[WW79]   A.G. Worthington and P.T. Wong. Effects of earned and assigned grades on student evaluations of an instructor. *Journal of Educational Psychology*, 71:764–775, 1979.

Index